FROM AN ANTAGONISTIC TO A SYNERGISTIC PREDATOR PREY PERSPECTIVE

FROM AN ANTAGONISTIC TO A SYNERGISTIC PREDATOR PREY PERSPECTIVE

Bifurcations in Marine Ecosystems

TORE JOHANNESSEN
Institute of Marine Research, Norway
University of Cape Town, South Africa

AMSTERDAM • BOSTON • HEIDELBERG • LONDON
NEW YORK • OXFORD • PARIS • SAN DIEGO
SAN FRANCISCO • SINGAPORE • SYDNEY • TOKYO
Academic Press is an imprint of Elsevier

Academic Press is an imprint of Elsevier
32 Jamestown Road, London NW1 7BY, UK
225 Wyman Street, Waltham, MA 02451, USA
525 B Street, Suite 1800, San Diego, CA 92101-4495, USA

British Library Cataloguing-in-Publication Data
A catalogue record for this book is available from the British Library

Library of Congress Cataloging-in-Publication Data
A catalog record for this book is available from the Library of Congress

ISBN: 978-0-12-417016-2

For information on all Academic Press publications
visit our website at elsevierdirect.com

Cover Photo: Øystein Paulsen, Institute of Marine Research

Dedication

To Aadne Sollie

Contents

List of Contributors

Ingrid Berthinussen Norwegian Polar Institute, Tromsø, Norway
Anders Fernö University of Bergen, Department of Biology, Bergen, Norway
Tore Johannessen Institute of Marine Research, Flødevigen, Norway
Jens-Otto Krakstad Institute of Marine Research, Bergen, Norway

Preface

In a field of science in which there is no confirmed theory, two approaches are needed for the advancement of the field: (1) existing perspectives should be tested and either rejected if not supported by data, or else, sufficient evidence should be provided to lift ideas from the status of hypotheses to a confirmed theory, and (2) existing perspectives should be challenged by new ideas. In my opinion, both approaches are equally important. Despite the fact that Darwin (1859) published his ideas on evolution 155 years ago and that these ideas have been developed into a well-established theory, the closely related field of ecology appears to be still in its infancy, with meager prospects of rapid advancement at the present rate of obtaining insight into ecological mechanisms. Much published work on ecology revolves around testing existing perspectives, of which many are conflicting, for example, global stability versus multiple stable points, stability in simple versus more complex systems, bottom-up versus top-down control, balance versus non-balance of nature, gradual dose-response relationships versus abrupt regime shifts. None of these opposing views have been resolved. Hence, there are few ideas in ecology that are embraced with consensus.

In the present dismal situation, new and challenging perspectives should be highly welcomed, not as facts but as ideas for the potential advancement of ecological theory. Indeed, if the existing perspectives do not provide a sufficient basis for developing a holistic ecological theory, new ideas are prerequisites for the advancement of the field. Over the years the ecological discussion has mainly revolved around the previously mentioned dichotomies, and in relation to this discussion, Hunter and Price (1992, p. 724) state: "there is a collection of idiosyncratic systems, with their associated protagonists, in which opposing views on the importance of various factors are debated," and that "one reason that opposing views are long-standing in the literature is that authors carry with them experiences and prejudice developed from the particular organisms that they study." Unfortunately, in combination with the review system of scientific journals, such "prejudice" appears to severely hamper the publishing of new and challenging ideas. Furthermore, project proposals are generally subjected to a similar review system, with meager prospects for provocative ideas to get financial support.

As ecological theory appears to be in its infancy, I believe that open-mindedness in combination with a sound critical attitude are important for the advancement of the field. Fortunately, there are such open-minded scientists, two of whom reviewed the proposal of this book: Brian J. Rothschild and Trond Frede Thingstad. I am most grateful for their positive reviews that made it possible to present these novel perspectives in ecology.

The ideas presented in this book were developed over a period of more than 20 years. During most of this time I was responsible for two fish stocks in the North Sea, lesser sandeel (*Ammodytes marinus*) and Norway pout (*Trisopterus esmarkii*), which took most of my time. Much of the development of these ideas took place during my stay at the University of Cape Town, South Africa, in 2002. I am most grateful to John G. Field for hosting me and for the kind reception I received from the staff of the Department of Zoology. This book is therefore a combined publication of the University of Cape Town and my employer, the Institute of Marine Research in Norway.

There are a number of people that have contributed to make this book much better than I would be able to do on my own. I am particular grateful to my dear colleague Odd Aksel Bergstad, who has commented extensively on all parts of the manuscript—thank you very much Odd Aksel! Thanks are also due to Petter Baardsen, Geir Huse, Espen Johnsen, Stuart Larsen, Lars J. Naustvoll and Oddvar Nesse, for their invaluable contributions during the writing process. The ideas and conclusions presented in this book are, however, entirely my responsibility, along with my coauthors of Chapter 4.

An essential pillar of this book is some near-century-long time series collected from the south coast of Norway by former and present colleagues at the Institute of Marine Research, Flødevigen. An annual beach seine sampling program was initiated by Alf Dannevig, former director at Flødevigen, in 1919. The field work was led by Rangvald Løversen, 1919–1967; Aadne Sollie, 1968–2001; and Øystein Paulsen, 2002–2012. Due to their enthusiastic and conscientious work, this survey has become probably one of the finest time series from marine ecosystems worldwide. It has been a privilege to analyze these fine data. I am very grateful to the three survey leaders and all those who have taken part in collecting data to this and other time series analyzed in this book, and to the Directorate of Nature Management that financed the transfer of the historical data from hand-written notes to digital formats. Thanks are also due to Odd Lindahl, University of Gothenburg, for generously providing data from his unique time series on primary productivity. Lastly, I am glad to thank my wife, Ingjerd Hompland, and my two sons, Ivar and Tord Hompland, for supporting me, listening to me, and ignoring me precisely as required.

References

Darwin, C., 1859. On the Origin of Species. Murray, London.
Hunter, M.D., Price, P.W., 1992. Playing chutes and ladders: heterogeneity and the relative roles of bottom-up and top-down forces in natural communities. Ecology 73, 723–732.

CHAPTER

1

Introduction

If at first the idea is not absurd, then there will be no hope for it.
Albert Einstein

1.1 ABOUT THIS BOOK

This book's title, *From an Antagonistic to a Synergistic Predator-Prey Perspective: Bifurcations in Marine Ecosystems,* includes two concepts that indicate the book's main focus, namely synergism in relation to predator-prey interactions and ecosystem bifurcations. In addition, recruitment variability in fish is dealt with in detail. Throughout the book, the terms *predator* and *prey* are used in the broadest sense of the words, including grazers as predators on primary producers. Predator-prey synergism, introduced as a new concept, is defined as predator-prey relationships enhancing abundances of both predator and prey. Hence, synergistic predators have a positive impact on the abundance of their prey, whereas antagonistic predators have negative impact on their prey. The idea of predator-prey synergism (hereafter "synergism") emerged as an alternative predator-prey model to account for phenomena observed in long-term time series (since 1919) from the south coast of Norway that appeared paradoxical under an antagonistic predator-prey model, for example, dominance of edible phytoplankton under high grazing pressure, red tides occurring in apparently nutrient depleted water, and repeated observation of bifurcations.

Ecosystem bifurcation is defined as an abrupt and persistent regime shift that affects several trophic levels and results from gradual environmental changes. The concepts and theoretical background for bifurcations are reviewed in this introduction. The empirical basis for suggesting that marine ecosystems are vulnerable to bifurcations is the previously mentioned long time series from the south coast of Norway,

From an Antagonistic to a Synergistic Predator Prey Perspective.
DOI: http://dx.doi.org/10.1016/B978-0-12-417016-2.00001-4

obtained during increasing anthropogenic eutrophication, and also increasing temperatures over the past 25 years.

This book consists of seven chapters plus this introduction. A full appreciation of the novel perspectives and theory will only be gained by reading the entire book. Each chapter was written to also satisfy readers wishing to read only selected chapters, however, and hence some repetition was unavoidable.

1.2 UNIFYING PRINCIPLES IN ECOLOGY—WHERE ARE WE?

At the dawn of the twentieth century, fishery biologists discovered that the year-class strength of many fishes varies substantially, and 100 years ago the Norwegian pioneer scientist Johan Hjort (1914) proposed the first recruitment hypothesis. This hypothesis suggests that recruitment variability results from different survival rates of fish larvae owing to the degree of match between the abundance of fish larvae and their prey. Hjort's hypothesis (or modifications of his hypothesis) has up until now been the most generally accepted explanation for recruitment variability (Houde, 2008). However, despite its having been one of the main focuses of marine research for more than 100 years, the recruitment puzzle remains unresolved. Being one of the most important structuring mechanisms in marine ecosystems, the lack of insight into what causes recruitment to vary is indeed compromising our present level of ecosystem knowledge. Unfortunately, this dismal situation appears to be not unique for marine ecosystems, but seems to reflect the generally rather low level of ecosystem understanding. There are few aspects in ecology that are embraced with general consensus, and Ridley (2003, p. 5) is probably right in stating that "[evolution] is the only theory that can seriously claim to unify biology."

One important approach of gaining insight into ecosystem mechanisms is by theoretical studies and mathematical modeling. Such studies show that it is theoretically possible that complex ecosystems are less stable than simple ecosystems (e.g., May, 1973). This was indeed an interesting perspective, as it turned the previous notion of ecosystem stability (Elton, 1958; MacArthur, 1955) upside down. However, the theoretical proposals of ecosystem mechanisms derived by mathematical models are per se, nothing but elegantly designed theoretical speculations from which to obtain new ideas that must be verified by studies of real ecosystems. Similarly, experimental ecosystem studies are inevitably too limited in terms of the number of species and spatial and temporal scales to be realistic (Pimm, 1991). Hence, both mathematical models and most experimental ecological studies can only provide ideas that will have to be tested by studies of real ecosystems on realistic temporal and spatial scales.

The problem faced in studying real ecosystems seems to be similar to that of studying evolutionary processes, which was so simply and convincingly formulated by Maynard Smith (1977, p. 236): "For example, consider selection. Suppose that there are two types of individuals in a population, say red and blue, which differ by 0.1 per cent, or 1 part in 1000 in their chances of surviving to breed. If the population is reasonably large (in fact, greater than 1000), this difference in chance will determine the direction of evolution, towards red or blue as the case may be. But if we wished to demonstrate a difference in the probability of survival during one generation—that is, wished to demonstrate natural selection—we would have to follow the fate of one million individuals, usually an impossible task."

Nevertheless, evolution is a solidly grounded theory for several reasons (Ridley, 2003), two of which may provide guidelines for the advancement of ecosystem theory: (1) Evidence of evolution in the fossil record (the temporal problem) and (2) a mechanistic explanation for evolution in terms of mutations and natural selection. One way to overcome the temporal problem in studies of natural ecosystems is to collect long and systematic time series and then, based on patterns in such time series, develop mechanistic models for the processes underlying the observed patterns. These models should then be tested, preferably in real ecosystems. Through observations from real ecosystems, we can be relatively certain that the studied phenomena are genuine ecosystem responses.

This approach was adopted in this book, which starts off by presenting results from systematically collected annual abundance data of young-of-the-year gadoid fishes along the south coast of Norway (since 1919). This time series revealed repeated incidents of sudden and persistent recruitment failures in the gadoids. Comprehensive testing in the field of the mechanism underlying these recruitment failures, and direct and indirect evidence of concurrent shifts in the plankton community, provided substantial evidence suggesting that marine ecosystems are vulnerable to bifurcations.

1.3 RECRUITMENT VARIABILITY

In order to disentangle the mechanisms underlying the recruitment failures in the gadoids, a recruitment hypothesis was suggested and tested in the field. Atlantic cod (*Gadus morhua*) was used as a model species in these studies. The generally accepted perception is that most of the recruitment variability in fish occurs during early life stages.

In agreement with this, cod recruitment was mainly determined within the first six months after spawning but well after the larval stage. The results suggested that young-of-the-year cod depend on energy-rich planktonic prey until they are quite large (up to approximately 8 cm), and early shifts to less energy-rich prey (e.g., fish and prawns) result in low condition and poor survival. It was proposed that variability in the plankton community generates variable energy flow patterns to higher trophic levels and thereby induces recruitment fluctuations in cod, other fishes, and benthic invertebrates that depend on pelagic prey during early life stages. After this period of food-limited survival, abundant organisms will attract opportunistic predators, which will then act to reduce differences between year-classes at older stages. It is suggested that, as general phenomena, physical and chemical bottom-up processes generate variability in marine pelagic food webs, whereas predation, parasitism, and diseases act to dampen variability. Fisheries targeting larger fishes will thus induce variability in marine ecosystems.

The mechanism underlying the repeated incidents of sudden and persistent recruitment failure in the gadoid fishes was suggested to be abrupt shifts in the plankton community as a result of gradual environmental changes (eutrophication and increasing temperature).

1.4 ECOSYSTEM BIFURCATION

With the prospect of global warming, a significant topical question is how marine ecosystem will respond to gradual environmental changes. At present, the assessment and monitoring of dynamics of ecosystems are based on the assumption of simple dose-response relationships. Gradual environmental changes or perturbations are expected to cause corresponding changes in the abundance of affected species. However, it has long been recognized theoretically that ecosystems may shift between alternative stable states, each of which has its own basin of attraction (Holling, 1973; Lewontin, 1969; May, 1977).

More recently, evidence of shifts between contrasting states in large-scale ecosystems was provided (Scheffer et al., 2001). Most of these examples were ecosystem shifts attributed to abrupt environmental shifts or catastrophic events (e.g., storms, mass mortality due to pathogens). One example, however, was the gradually increasing eutrophication in shallow lakes, causing shifts from a clear water state with submerged vegetation to a turbid state in which phytoplankton dominated. Such shifts have been classified as bifurcations (Biggs et al., 2009; Scheffer et al., 2009). According to mathematical theory, a bifurcation occurs when a small, smooth change made to the parameter values of a system causes a sudden change in system behavior.

In the marine literature, the term "regime shift" has been frequently used to describe abrupt changes in time series. However, the application of different definitions of regime shifts (Jarre et al., 2006; Overland et al., 2008) has rendered the concept vague. To be more explicit, the concept of ecosystem bifurcations used in this book refers to abrupt and persistent ecosystem shifts that affect several trophic levels and result from gradual environmental changes. To define more precisely, the term "persistent" is not straightforward. A useful criterion, though, could be the one suggested by Connell and Sausa (1983, p. 808) to judge whether real ecosystems are stable: "the fate of all adults must. . .be followed for at least one complete overturn. . . ."

Resilience is a concept inseparably linked to ecosystem bifurcations. However, there are different definitions of the concept (Gunderson, 2000; Pimm, 1991). Here, resilience is used as proposed by Holling (1973), defined as the maximum perturbation a system can sustain without causing a shift to an alternative stable state.

Ecosystem bifurcations are not restricted to shifts triggered when tipping points in critical variables are reached. Bifurcations may also occur when shifts are triggered by environmental perturbations after the resilience of the system has been reduced as a result of gradual environmental changes, that is, the shift may occur before the tipping point is reached.

The theoretical relationships of bifurcations, resilience, and environmental perturbations is illustrated in Fig. 1.1 (modified from ideas by Lewontin, 1969; May, 1977; Scheffer et al., 2001). In nature, there are different dynamically stable ecosystem states—that is, the community structure varies within specific limits. A stable state can be considered as a trough in which a ball is being rocked back and forth by environmental and biological perturbations (e.g., temperature variability, diseases, and invasions). The depth of the trough represents resilience. Under a specific environmental regime the ecosystem state for which the conditions are optimal will have the highest resilience (Fig. 1.1a and c). When the resilience is high, large perturbations are needed to bring the ball out of the trough. If the environmental conditions change in favor of State 2 (Fig. 1.1b), the resilience of State 1 will be reduced and the ecosystem may become vulnerable to bifurcation from perturbations the system normally could withstand. Perturbations may, for example, be in the form of mass mortality caused by diseases, toxic algal blooms, or low or high temperatures, which may pave the way for organisms that are better adapted to the altered environmental regime. Hence, under this interpretation of the mechanism underlying bifurcations, the organisms with competitive advantages will dominate the system after a severe perturbation.

The concept of bifurcations implies that stepwise changes are considered gradual if the response curve is abrupt and the initial steps do not

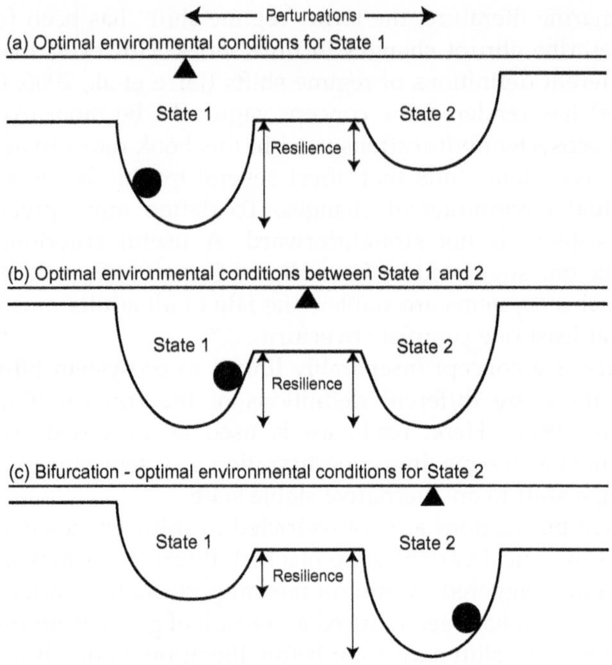

FIGURE 1.1 Conceptual model for the relationships of resilience, changing environmental conditions, and perturbations: (a) optimal conditions for State 1, (b) optimal conditions between State 1 and State 2, (c) ecosystem bifurcation, optimal conditions for State 2.

cause an ecosystem shift. If an abrupt shift in the environmental conditions elicits a concurrent shift in the ecosystem, it would not be classified as a bifurcation because this would suggest a simple dose-response relationship. On the other hand, a shift in the environmental conditions may reduce the resilience of the ecosystem without causing an immediate shift. If the ecosystem later shifts because of an environmental or biological perturbation, it would be considered a bifurcation.

1.5 PREDATOR-PREY SYNERGISM

There is growing evidence from both aquatic and terrestrial ecosystems that the relationships between primary producers and herbivores are complex and include both the direct impact of grazing and the indirect impact of the recycling of nutrients (Elser and Urabe, 1999; McNaughton et al., 1997; Sterner, 1986). Nevertheless, it is still an important assumption in ecological theory that interactions between predators and prey are mainly antagonistic (Loreau, 1995), implying that a high abundance of a predator reduces the abundance of its prey. From the perspective of systems ecology, this assumption is linked to the perception that nutritional requirements cause organisms to compete for resources and/or to feed on each other, leading to negative interactions between populations (competition, predation, parasitism), with symbiosis as a rather exotic case (Sommer, 1989). Also, from a reductionist perspective (the level of the individual) the predator-prey relationship is obviously negative because the predator either kills or damages its prey. A potential problem with both of these perspectives is that only the direct predator-prey relationship is considered. It is conceivable that even though a grazer has a negative impact on its preferred plants seen in comparison with specimens of the same species that are not being grazed, the preferred plants may gain competitive advantages over non-grazed species by being only mildly affected by the grazing and by the grazer weeding out non-grazed competitors (e.g., by clipping sprouts of tall vegetation). McNaughton (1979) calls this competitive fitness.

This book presents data from a marine spring blooming ecosystem, which suggests that such positive relationships do exist between phytoplankton and herbivorous zooplankton. A model for the interactions between zooplankton and their algal prey is proposed, which provides a mechanistic explanation for the competitive benefits of grazed algae compared to non-grazed competitors. The model does not rely on group selection but is in accord with the perception of "the selfish gene." It is also argued that synergism is an evolutionary stable strategy that is, it cannot be invaded by a mutation resulting in non-palatability. By taking this positive relationship between zooplankton and their preferred algal prey into account, it is further suggested that the planktivores at the next trophic level also have a positive impact on the abundance of their prey. Hence, synergism appears to comprise the three lowest trophic levels of the marine pelagic food web. However, there are alternative strategies to synergism that are considered as loopholes in a synergistic realm. Synergistic interaction in the pelagic food web potentially represents a new paradigm in aquatic ecology and as such is a building block for revising ecological theories.

Potential implications of synergism are investigated, including annual succession patterns in the plankton community and processes underlying declining resilience that render marine ecosystems vulnerable to bifurcations. Synergistic relationships may have wide-ranging implications for the production and harvesting potential of marine resources and thus for fisheries management. Gradual global warming has the potential to induce bifurcations, resulting in substantially reduced production of fish due to recruitment failure. Reduced fish abundances will in turn reinforce the negative impacts of global warming and possibly render the ecosystem shifts irreversible. Management measures mitigating such negative impacts of global warming are proposed. It is also suggested that that synergism may be important in some terrestrial systems.

Resilience in marine ecosystems may be explained mechanistically by synergism, as may the repeated incidents of bifurcations observed in the long-term time series. Synergism generates substantial temporal dependency (autocorrelation) in marine ecosystems and thus partly overrides and modifies impacts of the bottom-up process from physical and chemical variability. In ecosystems in which synergistic interactions are important, there are no simple dose-response relationships, and most correlation studies between bottom-up processes and biological responses are therefore doomed to fail.

References

Biggs, R., Carpenter, S.R., Brock, W.A., 2009. Turning back from the brink: detecting an impending regime shift in time to avert it. Proc. Natl. Acad. Sci. USA. 106, 826–831.

Connell, H., Sausa, W.P., 1983. On the evidence needed to judge ecological stability or persistence. Am. Nat. 121, 789–824.

Elser, J.J., Urabe, J., 1999. The stoichiometry of consumer-driven nutrient recycling: theory, observations, and consequences. Ecology 80, 745–751.

Elton, C.S., 1958. The Ecology of Invasions by Animals and Plants. Chapman and Hall, London.

Gunderson, L.H., 2000. Ecological resilience—in theory and application. Annu. Rev. Ecol. Syst. 31, 425–439.

Hjort, J., 1914. Fluctuations in the great fisheries of northern Europe viewed in the light of biological research. Rapp. P.-v. Réun. Cons. int. Explor. Mer 20, 1–228.

Holling, C.S., 1973. Resilience and stability of ecological systems. Annu. Rev. Ecol. Syst. 4, 385–398.

Houde, E.D., 2008. Emerging from Hjort's shadow. J. Northw. Atl. Fish. Sci. 41, 53–70.

Jarre, A., Moloney, C.L., Shannon, L.J., Fréon, P., van der Lingen, C.D., Verheye, H.M., et al., 2006. Developing a basis for detecting and predicting long-term ecosystem changes. In: Shannon, V., Hempel, G., Malanotte-Rizzoli, P., Moloney, C., Woods, J. (Eds.), Benguela: Predicting a Large Marine Ecosystem. Large Marine Ecosystem. Elsevier, Amsterdam, pp. 239–272, Series 14.

Lewontin, R.C., 1969. The meaning of stability. Brookhaven Symp. Biol. 22, 13–24.

Loreau, M., 1995. Consumers as maximizers of matter and energy flow in ecosystems. Am. Nat. 145, 22–42.

MacArthur, R.H., 1955. Fluctuations of animal populations and a measure of community stability. Ecology 36, 533–536.

May, R.M., 1973. Stability and Complexity in Model Ecosystems. Princeton University Press, Princeton.

May, R.M., 1977. Thresholds and breakpoints in ecosystems with a multiplicity of stable states. Nature 269, 471–477.

Maynard Smith, J., 1977. The limitation of evolution theory. In: Duncan, R., Weston-Smith, M. (Eds.), The Encyclopedia of Ignorance: Life Sciences and Earth Sciences. Pergamon Press, Oxford, pp. 235–242.

McNaughton, S.J., 1979. Grazing as an optimization process: grass-ungulate relationships in the Serengeti. Am. Nat. 113, 691–703.

McNaughton, S.J., Banyikwa, F.F., McNaughton, M.M., 1997. Promotion of the cycling of diet-enhancing nutrients by African grazers. Science 278, 1798–1800.

Overland, J., Rodionov, S., Minobe, S., Bond, N., 2008. North Pacific regime shifts: Definitions, issues and recent transitions. Prog. Oceanogr. 77, 92–102.

Pimm, S.L., 1991. The Balance of Nature? University of Chicago Press, London.

Ridley, M., 2003. Evolution. Blackwell Scientific Publications, London.

Scheffer, M., Carpenter, S., Foley, J., Folke, C., Walker, B., 2001. Catastrophic shifts in ecosystems. Nature 413, 591–596.

Scheffer, M., Bascompte, J., Brock, W.A., Brovkin, V., Carpenter, S.R., Dakos, V., et al., 2009. Early-warning signals for critical transitions. Nature 461, 53–59.

Sommer, U., 1989. Toward a Darwinian ecology of plankton. In: Sommer, U. (Ed.), Plankton Ecology: Succession in Plankton Communities. Springer-Verlag, Berlin, pp. 1–8.

Sterner, R.W., 1986. Herbivores' direct and indirect effects on algal populations. Science 231, 605–607.

CHAPTER

2

Repeated Incidents of Abrupt and Persistent Recruitment Failures in Gadoids in Relation to Increasing Eutrophication, 1919–2001

2.1 INTRODUCTION

The technological development in the twentieth century was formidable. In industrialized countries, this led to substantial improvements in the standard of living. The cost of such development was increased pressure on nature and ecosystem processes. Until the 1960s to 1970s there was little concern about the environmental impact of the technological revolution. Typically for this early period, leading waste products through pipes into the sea was considered a practical way of getting rid of refuse. In addition, every pollutant, whether in the air or on land, tended to end up in the ocean (Williams, 1996). Hence, the lack of concern regarding the impact of pollution led to substantial nutrient loads in coastal waters in industrialized and densely populated areas. Nutrients came from municipal sewage plants as water closets gradually replaced dry closets, increasing use of synthetic fertilizers in agriculture, industrial point sources and atmospheric deposition of nutrients from combustion of fossil fuels. In many places, coastal waters were also contaminated by heavy metals (e.g., mercury, lead, cadmium, chrome, and arsenic) and synthetic chemicals (such as the insecticide DDT and the chemical compounds of PCB that were widely used in a variety of electrical appliances). In general, the various effluents increased gradually because technological

From an Antagonistic to a Synergistic Predator Prey Perspective.
DOI: http://dx.doi.org/10.1016/B978-0-12-417016-2.00002-6

development and new technological achievements are mostly implemented gradually in society. For example, in the Norwegian capital Oslo water closets increased gradually from only few in 1900 to approximately 80,000 in 1940 (http://www.vann-og-avlopsetaten.oslo.kommune.no/getfile.php/vann-%20og%20avløpsetaten%20(VAV)/Internett%20(VAV)/Dokumenter/studie/historie/kap_5.pdf), and the use of synthetic fertilizers in Norway gradually tripled from 1950 to 1980 (http://www.ssb.no/natur-og-miljo/artikler-og-publikasjoner/gjodsel-ressurs-men-miljoproblem).

In the 1960s and 1970s there was growing concern of the environmental impact of pollution. In Norway, this led to the prohibition of DDT in 1970 and PCB usage in 1980 (http://www.environment.no). Furthermore, considerable effort was made to reduce or stop contamination of nature by heavy metals and other harmful chemicals and to reduce nutrient loads from municipal sewage plants and industrial point sources.

The increasing nutrient loads and contamination of coastal waters in the twentieth century followed by reductions of many of these effluents can be regarded as large-scale experiments on real ecosystems. Numerous studies have documented substantial changes and damage in ecosystems in coastal waters and enclosed and semi-enclosed seas as result of pollution (Caddy, 1993; Islam and Tanaka, 2004). Unfortunately, prior to the growing environmental awareness in the 1960s and 1970s there was very little monitoring of the ecological impact of pollution. Hence, the shape of the dose-response relationship of pollution-induced changes in coastal ecosystems is often inadequately documented.

Time series on the abundance of marine organisms obtained in coastal waters during the course of pollution may potentially be used to assess both the ecological impacts and the shape of the dose-response relationship, presuming the data have been obtained systematically. Long-term time series (greater than 50 years) are in general rare, and long-term time series without significant methodological changes are exceptional. This chapter describes historical variability in the abundance of 0-group (young-of-the-year) gadoids from the Norwegian Skagerrak coast (Fig. 2.1) based on one such exceptional unbroken time series—an annual beach seine survey in which the sampling methodology has remained unchanged since the start in 1919. Not only were the methods and locations sampled practically unchanged through the series, but until 2001 only two persons had run the fieldwork, and they had an overlap of 10 years!

The period encompassed by the data sets used in this study (1919 to 2001) includes the major increases of anthropogenic effluents in

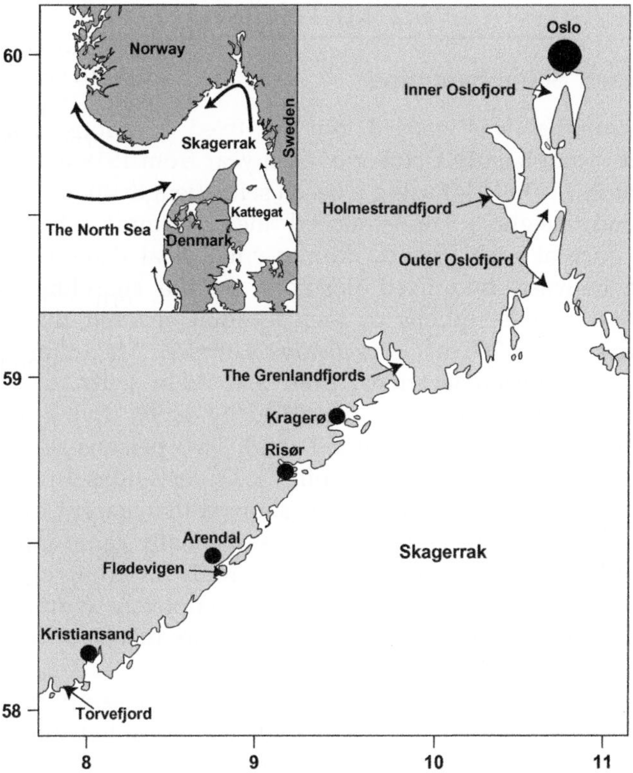

FIGURE 2.1 Beach seine sampling areas along the Norwegian Skagerrak coast.

Skagerrak coastal waters (results after 2001 are presented in Chapter 5 due to different ecosystem responses; Johannessen et al., 2012). The Norwegian Skagerrak coast is relatively densely populated and includes the capital Oslo, with a present population of about 750,000 surrounding the semi-enclosed Inner Oslofjord (Fig. 2.1), and some heavily industrialized areas. In addition to local sources, pollutants (including nutrients) from the southern North Sea, Kattegat, and the Baltic may affect the environment as most of the water masses from these areas pass along the Norwegian Skagerrak coast to waters farther north (Aure et al., 1998). Corresponding to the increase in pollution during the twentieth century, the environment of the coastal waters of Skagerrak has been characterized by marked declines in oxygen concentrations of both intermediate water masses and bottom waters (Johannessen and Dahl, 1996a, 1996b).

2.2 METHODS

2.2.1 Beach Seine Sampling

Fish sampling was carried out at fixed locations during late September through early October of each year from 1919 (except during World War 2, 1940–1944) using a beach seine (length 40 m + 30 m ropes at each end, height 1.7 m, stretched mesh 15 mm). The seine was deployed once at each location from a small boat (5.6 m) rowed in a semicircle between two fixed positions on the shoreline (see front cover). Detailed descriptions of each location ensured that the same bottom area (up to 700 m^2) was always sampled. Maximum sampling depth varied between locations, from 3 to 17 m with an average of 7.1 m (average of 117 locations described by scuba diving and use of eco-sounder; Skagerrak is a nontidal area). Two persons slowly hauled up the seine (approximately 3 m min^{-1}). Observations by divers and from the surface using an aqua-scope showed that, except for flatfishes that rest on the bottom and therefore occasionally came into physical contact with the seine, the fish swam calmly in front of the slow-moving net. Only after having become completely entrapped were panic and attempts to escape observed (for more details about gear and sampling, see Tveite, 1971).

The 0-group gadoids were the target fishes of the sampling program. For each location, the catch of the gadoids were counted, and up to 100 individuals per location and species were measured to the closest centimeter. Before 1970 the length of all gadoids were measured. Average catch per beach seine haul was used as an annual abundance index for five 0-group gadoid species: cod (*Gadus morhua*), pollack (*Pallachius pollachius*), whiting (*Merlangius merlangus*), saithe (*Pollachius virens*), and poor-cod (*Trisopterus minutus*). The number of length measurements of poor-cod was insufficient in some years to distinguish between 0-group and older year-classes. For poor-cod the abundance indices therefore comprise all year-classes, but 0-group usually dominated by contributing to 79% and I-group to 21% of the total number of poor-cod (estimated for the period 1989–2001). The abundance time series were smoothed by computing the 7-year moving average twice (i.e., smoothing the raw time series and then smoothing the smoothed time series, which corresponds to a locally weighted 13-year moving average where the central value is included seven times, the values on both sides of the central value six times, the following values five times, etc.), which gives a smooth curve with much local weight.

Before 1989, the catch of all other fishes and some invertebrates were either counted or recorded semiquantitatively (null, few, some, many, and numerous). After 1989, all fishes have been counted and measured by length. Some results obtained after 1989 on these species have been presented in this chapter.

2.2.2 Bottom Vegetation

In the early 1930s there was extensive mortality of eelgrass (*Zostera marina*) throughout the North Atlantic (Short et al., 1988). According to logbook notes by cruise leader Ragnvald Løversen, the decimation of eelgrass occurred in 1933 along the Norwegian Skagerrak coast. The level before 1933 is unknown, but according to Løversen's descriptions, the reduction in eelgrass cover must have been substantial. Due to this event, bottom flora coverage as observed from the surface using an aqua-scope was recorded from 1934 onwards. The aqua-scope was a 60-cm-long conical cylinder with a black innerwall and equipped with a 25-cm-diameter glass disk in the wide end and face-adaptation in the narrow end. From these observations, an index of the bottom flora (mainly eelgrass, but also benthic macroalgae) coverage was estimated on a relatively coarse scale. Because these observations are semiquantitative, and poor visibility in some years made observations difficult, focus is on trend rather than individual years.

2.2.3 Study Sites

Presently, approximately 130 beach seine locations are included in the sampling program, of which 38 have been sampled consistently since 1919. In this chapter, general trends in 0-group gadoids on the Norwegian Skagerrak coast are described on the basis of these 38 locations alone. This subset of locations represents the coastline between Torvefjord west of Kristiansand and Kragerø (Fig. 2.1).

Another subset of locations was selected to describe and compare species-specific distributional differences related to distance to the open coastline, that is, locations in the area Sandnesfjord–Risør archipelago (Fig. 2.2a). This area was chosen because Sandnesfjord is a topographically quite simple straight fjord with the Risør islet archipelago just outside. In addition, except for a river entering the head of Sandnesfjord, draining woodland areas with low human population densities, there are no major local effluents affecting the environment. By 1993, the oxygen saturation in Sandnesfjord was estimated at 90, 79, and 41% at depths of 10, 30, and 60 m (bottom), respectively (Johannessen and Dahl, 1996a), indicating generally good oxygen conditions and relatively high water exchange rates with offshore neritic water masses.

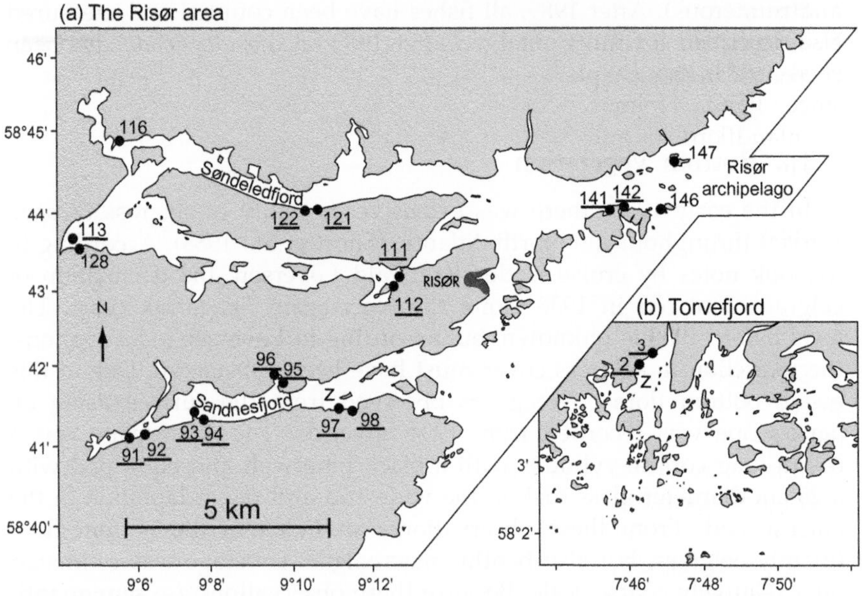

FIGURE 2.2 Beach seine sampling locations in (a) the Risør area and (b) Torvefjord. Underlined locations have been sampled since 1919 and are included in this chapter, and the other locations are included in a study presented in Chapter 4. Z indicates position of hydrographical stations.

Differences between exposed and semi-enclosed areas were described for Torvefjord (Fig. 2.2b) and inner Søndeledfjord (Fig. 2.2a). The relatively exposed Torvefjord is the most westerly sampling area, and the two locations sampled are situated in the inner part of an archipelago, approximately 5 km from the open coastline. The two locations in the semi-enclosed Søndeledfjord are situated in the head of the U-shaped fjord, approximately 20 km from the open coastline. Except for limited outflow of fresh water from small rivers or brooks, neither area has significant local effluents affecting the environment. Despite relatively low fresh water supply, the salinity of the surface water in Søndeledfjord is generally low (approximately 17 PSU at 1 m depth, avg. 1994–1999 in September), reflecting low water exchange with coastal water. For the same reason, oxygen concentrations are low, with close to permanent anoxia below 20–30 m (Johannessen and Dahl, 1996a). Surface water salinity in Torvefjord is much higher (approximately 27 PSU, avg. 1994–1999 in September), and oxygen conditions at 30 m depth are generally good with ≥ 80% saturation (1994–1999 in September). Hence, water exchange with offshore neritic water masses is much higher in Torvefjord than in Søndeledfjord.

Historical changes in the abundance of 0-group gadoids were also described for three polluted areas (Fig. 2.1; e.g., Alve, 1995; Bayne et al., 1988; Ruud, 1968a): Inner Oslofjord (nine beach seine locations since 1936), Holmestrandfjord (six locations since 1936), and the Grenlandfjords (seven locations since 1953).

The exact position of all sampling locations included in this chapter can be seen in Figures 45–54 in Johannessen and Sollie (1994; http://brage.bibsys.no/imr/handle/URN:NBN:no-bibsys_brage_3791).

2.2.4 Sampling Reliability

To assess the reliability of the beach seine data, the abundance indices of 0-group were correlated with the abundance indices of I-group the following year (i.e., the same year-class). All beach seine locations sampled over periods in excess of 20 years were included (1919–2001, average 81 locations per year), except locations from the three polluted areas mentioned above. The 0- and I-group cohorts were separated using length measurements. Whiting and poor-cod were excluded from this analysis on the basis of negligible catches of whiting older than one year, and too poor length data for poor-cod to allow separation into age-groups on an annual basis. The overlap between 0-, I-, and II-group cohorts of both pollack and saithe was negligible, and the cohorts could consequently be readily separated on the basis of fish size. In the case of cod, fish of less than ≤15 cm were classified as 0-group, and fish from 20 to 30 cm were classified as I-group. Otolith readings have confirmed that 0-group cod can be up to 20 cm; I-group can be as small as 16 cm and a small proportion larger than 30 cm; a few II-group cod can be smaller than 30 cm. Because the majority of 0-group cod are smaller than 15 cm, the error in the length-based separation is negligible. The I-group index, on the other hand, can be underestimated by variable proportions of the fish being smaller or larger than the limits used here and overestimated in years with a strong year-class of II-group and a poor year-class of I-group. The correlation analyses were carried out on log transformed average catches; because zero values of pollack and saithe were observed, a value of 1 was added.

2.2.5 Statistical Analyses

The statistical method STARS (Rodionov and Overland, 2005) was used to detect abrupt shifts in time series. The method was run with a cut-off length of 10 (number of years of the regimes to be tested), target probability of 0.05, and Huber's weight parameter set to 5 to eliminate extreme outliers (set this high because very strong year-classes are normal in recruitment time series). After the detection of significant shifts in the time series, potential trends following such shifts were analyzed using simple linear regression when considered relevant. In some cases differences between mean catches during different periods were analyzed using the nonparametric Mann-Whitney U test.

2.3 RESULTS

The main focus is on 0-group cohorts of the most abundant gadoid species, namely cod, pollack, and whiting. Saithe and poor-cod have been included in the descriptions of general trends because they shed light on the mechanisms behind observed changes in gadoid abundance. Cod has received special attention because it is used as a model species in Chapters 3 and 4 in order to test these mechanisms.

2.3.1 Sampling Reliability

Because the reliability and precision of the beach seine sampling are essential for the remainder of this chapter, the results of these analyses are discussed here. For cod (Fig. 2.3) the coefficient of determination (r^2) between the beach seine abundances of the same year-classes measured at the 0-group and I-group stages in September was 0.72, for pollack 0.69, and for saithe 0.30, all of which were statistically significant

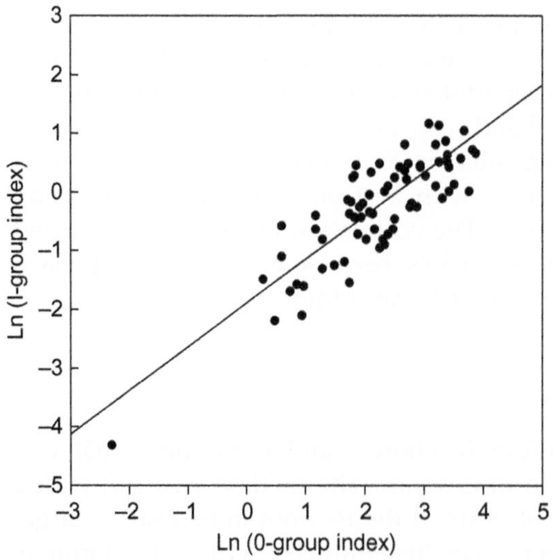

FIGURE 2.3 Relationship between year-class strength of cod at the 0-group and I-group stages. Year-class indices are log-transformed.

($p \ll 0.001$). Similarly, Tveite (1971, 1992) found positive correlations between the 0-group abundance of cod from the September beach seine survey and numerical catches of I–IV-group of the same year-class sampled in pots and in trammel nets.

Studies of coastal cod around the North-Atlantic have shown that the main distribution depth of 0-group cod is <10 m (Methven and Schneider, 1998; Riley and Parnell, 1984), which is in agreement with observations from the Skagerrak coast (authors's diving observations). The average maximum sampling depth of the beach seine at the various locations was 7.1 m (range 3–17 m). The seine therefore probably sampled the main depth range of the 0-group cod. The average abundance of I-group cod was merely 7.7% of that of the 0-group, whereas in pollack it was 18.6%. As the coastal cod grow older, they migrate to deeper water (Methven and Schneider, 1998; Riley and Parnell, 1984). Therefore, the reduction in abundance from the 0- to I-group stage cannot be used to estimate mortality. Low catch rates of I-group cod affect the estimate of r^2 between the abundance of 0- and I-group cod as it dropped from 0.72 to 0.56 when based on the subsample of the 38 locations sampled regularly since 1919. Furthermore, r^2 between the 0-group abundance indices of the total sample (average 81 locations) and the subsample was 0.95, but merely 0.60 between the I-group indices. These results suggest that the beach sampled the 0-group cod substantially more precisely than I-group cod, probably because the seine covered the main depth range of the 0-group cod.

On the other hand, based on a high number of locations, the r^2 between the 0- and I-group was high for both cod and pollack ($p \ll 0.001$). Obviously, such highly significant relationships cannot arise from noise and therefore suggest that the beach seine does sample the abundance of 0-group of both cod and pollack relatively precisely, and that also relatively precise estimates of the abundance of I-group cod and pollack can be obtained if the number of sampling locations are high. Furthermore, these results suggest that the year-class strength of both cod and pollack is mainly determined at the 0-group stage and prior to the survey period in September each year. Catches of 0-group and in particular I-group saithe are usually small (Fig. 2.4d). However, the positive correlation between 0- and I-group saithe suggests that the beach seine catches also reflect the approximate level of 0-group saithe abundance.

The calm behavior of the fish herded by the slow-moving seine (approximately 3 m min^{-1}) and the high correlations found in cod and pollack suggest that the beach seine is an adequate sampling tool for obtaining abundance indices of fishes inhabiting shallow water in general.

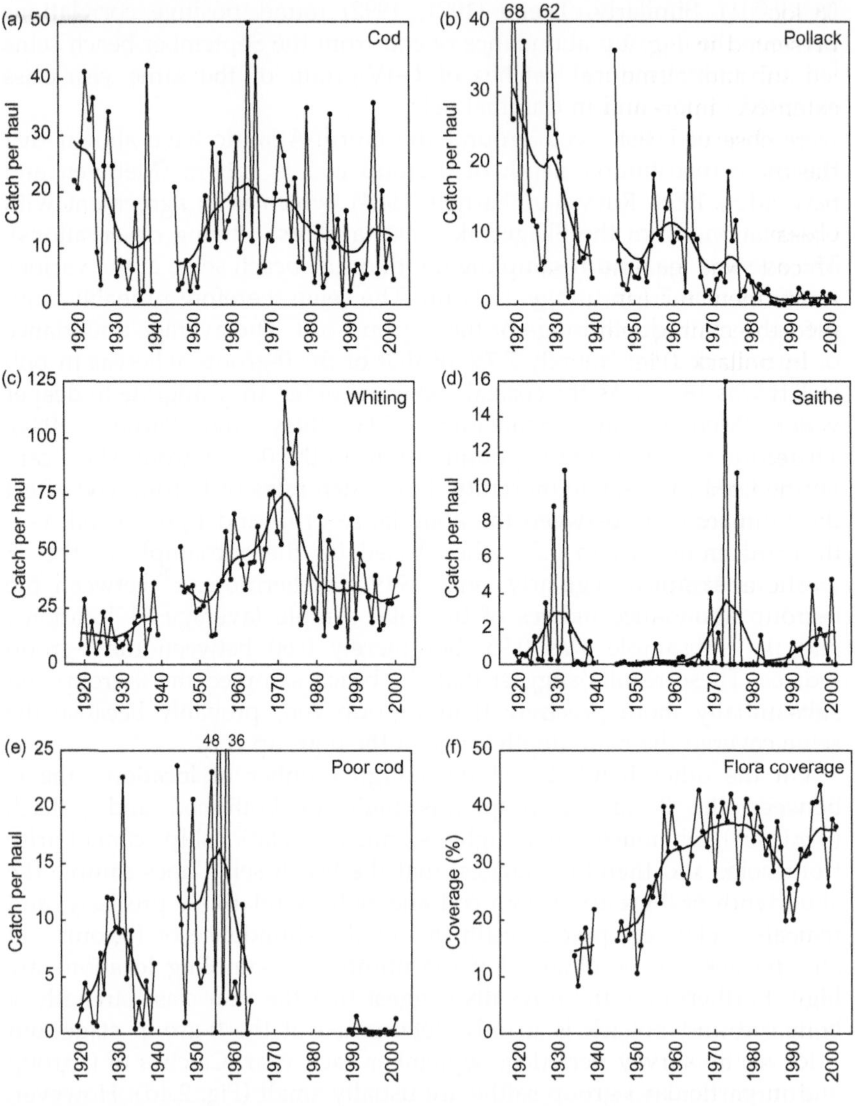

FIGURE 2.4 Average catch of 0-group (a) cod, (b) pollack (c) whiting, (d) saithe, and (e) poor cod (≥0-group) at 38 beach seine locations along the Skagerrak coast 1919–2001, and (f) bottom flora coverage at the beach seine locations 1934–2001. Smoothed curves correspond to a seven-year moving average estimated twice.

2.3.2 Temporal Variation of 0-Group Gadoids Abundance Along the Norwegian Skagerrak Coast

2.3.2.1 General Trends

The variation in beach seine abundance of 0-group gadoids and bottom flora coverage along the Skagerrak coast (i.e., the selected

38 locations along the coastline between Kristiansand and Kragerø) is presented in Fig. 2.4. For cod, pollack, and whiting the abundance varied substantially during the near-century-long time series. Despite the extensive inter-annual variability, some significant long-term trends were observed. For cod there were a number of successive strong year-classes during the 1920s, contrasting somewhat with the perception that two strong year-classes of cod rarely appear in succession (e.g., Gjøsæter and Danielssen, 1990). During the 1930s and 1940s recruitment of cod was generally poor, but from the early 1950s onwards the recruitment increased gradually to a relatively high level in the 1960s before declining again in the 1970s.

In pollack (Fig. 2.4b), the most conspicuous long-term trend was a steady decrease in abundance from the late 1920s onwards. The last year with reasonably high catches of 0-group pollack was 1976 and the subsequent average annual catch rate of pollack was only 4% of the average prior to 1930. Apart from this persistently much reduced abundance in recent decades, the pollack and cod abundances show much the same pattern with marked troughs in the 1930s and 1940s.

The abundance of 0-group whiting (Fig. 2.4c) showed an opposite trend to that of pollack until the mid-1970s, with increasing abundance from around 15 whiting per haul in the 1920s to around 100 in the 1970s. However, simultaneous with the drop in cod and pollack in the mid-1970s, there was an abrupt drop in whiting to around 30 fish per haul.

For saithe (Fig. 2.4d) there were two marked periods with high abundances, one in the 1920–1930s and another in the 1970s. In the 1990s there was also relatively high abundance. However, there is no historical trend in abundance of saithe ($p = 0.700$, linear regression).

As with pollack the recruitment of poor-cod has declined to what appears to be very low levels in recent decades (Fig. 2.4e). The time of the strong decline cannot be determined accurately as poor-cod was only recorded semiquantitatively between 1965 and 1988. The mean annual catches decreased by 96% between these periods (i.e., prior to 1965 and after 1988; $p < 0.001$, Mann-Whitney U test).

Bottom flora and particularly eelgrass meadows have long been recognized as important habitats for juvenile gadoids and other fishes (Blegvad, 1917). After the mass mortality of eelgrass in 1933, bottom flora coverage (Fig. 2.4f) increased from the 1930s to the 1960s and, except for a temporary decrease in the late 1980s, remained at this relatively high level until 2001. The increase of bottom flora coverage was mainly due to recovery of eelgrass.

Abrupt changes in the time series of cod and pollack occured in 1930 and 1932, respectively ($p \leq 0.006$, STARS), roughly but not perfectly matching the eelgrass mortality incidence in 1933. Perfect matches cannot be expected in time series in which the magnitude of some of the observations before and after shifts overlap; hence it is not unreasonable to interpret this observation as coincidence between fish abundance and decline in eelgrass coverage. Due to the rapid recovery of eelgrass, STARS detected a shift in the bottom flora time series in 1953 ($p < 0.001$). Cod and pollack abundance increased again during the recovery of bottom flora in the 1950s, and for cod a shift was detected in 1955 ($p = 0.024$). Shifts in the whiting time series were detected in 1957 ($p < 0.001$) and in 1976 ($p < 0.001$). Hence, the upward shifts in both cod and whiting in the mid-1950s matched the upward shift in bottom flora. The shift in whiting in 1976, however, appeared decoupled from the variation in bottom flora coverage. After the shift in 1976 there was no trend in whiting ($p = 0.807$, linear regression 1976–2001).

2.3.2.2 Spatial Variation at Selected Locations

Considering first variation perpendicular to the coastline (i.e., from the outer coast to inner fjord sites), Fig. 2.5 shows the abundance of 0-group cod, pollack and whiting plotted against the distance from sampling locations to the open coastline (the line connecting outer skerries). All data are from the subset of locations in the Sandnesfjord and Risør archipelago (Fig. 2.2). Each point in the main panels represents the average catch at two closely situated locations (Fig. 2.2a) over three 21-year periods (1919–1939—the first 21 years; 1955–1975—the period before the abrupt decrease in whiting; 1981–2001—the last 21 years). In addition, overall averages (average of all locations) for each period are included as subpanels.

The overall abundance of cod was about the same during the first and second period but approximately 50% lower during the last period. During all three periods, cod was generally more abundant in the head of the fjord than in locations nearer the coastal archipelago ($p < 0.001$ for periods one and two, $p = 0.023$ for period three, linear regression between distance and log-transformed individual catches, that is, the raw data). The overall decrease in abundance during the last period mainly occurred in the inner part of the fjord. A markedly asymmetric distribution of cod during the first two periods thus changed to a more even distribution.

The overall picture for pollack was a marked reduction from the initial to the middle period and a collapse in the final period. During the first period, pollack was almost absent in the inner fjord. In the middle period, all locations inside the fjord had low levels of abundance and pollack were relatively abundant only in the archipelago. During the last period, pollack became almost absent in the archipelago as well. In all three periods the abundance of pollack generally decreased with increasing distance from the open coastline ($p \leq 0.002$).

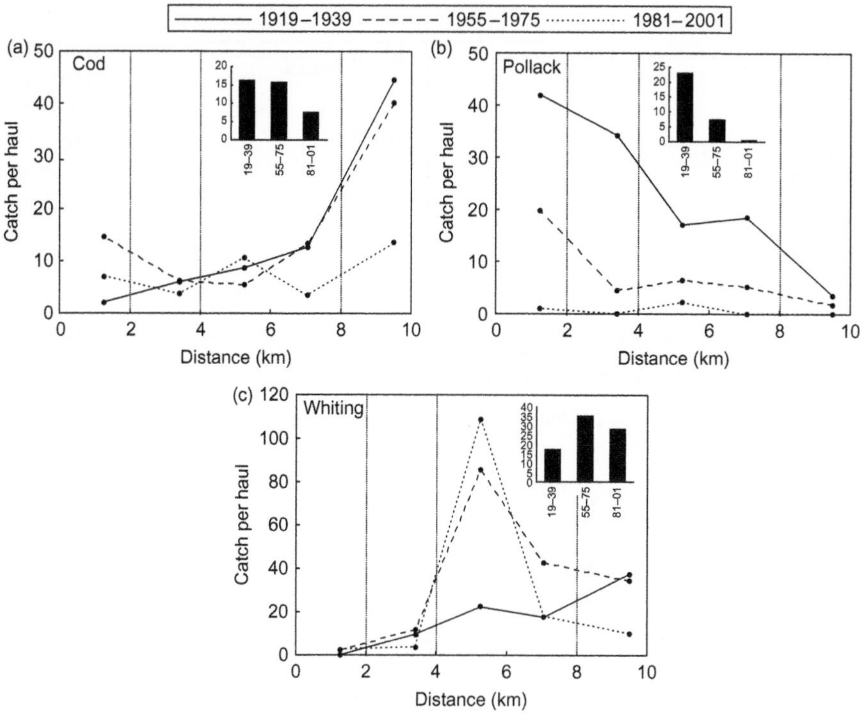

FIGURE 2.5 Average catch per haul of (a) cod, (b) pollack and (c) whiting in relation to distance to open coastline in Sandnesfjord and Risør archipelago (Fig. 2a). Each point represents the average at two closely situated locations during three 21-year periods, 1919–1939, 1955–1975, and 1981–2001. Vertical lines indicate archipelago (0–2 km), outer fjord (2–4 km), mid-fjord (4–8 km), and inner fjord (> 8 km). Bar graphs show the average catch for all locations during the various periods.

The abundance of whiting increased from the first to the second period, followed by a decrease during the last period. The distribution of whiting during the first period was similar to that of cod, with abundance increasing toward the inner fjord ($p < 0.001$). Increased abundance during the intermediate period was restricted to the mid-fjord, while the reduced abundance during the last period was observed in the inner fjord (≥7 km). Whiting showed consistently low levels in the archipelago and outer fjord.

During the last period, the two locations located 5 km from the open coastline (95 and 96; Fig. 2.2) did not follow the general trends of decreasing abundance for cod and whiting. In addition, during the last period the highest catches of pollack were taken at this location. The two locations are situated in an semi-enclosed cove. Location 96 had minor

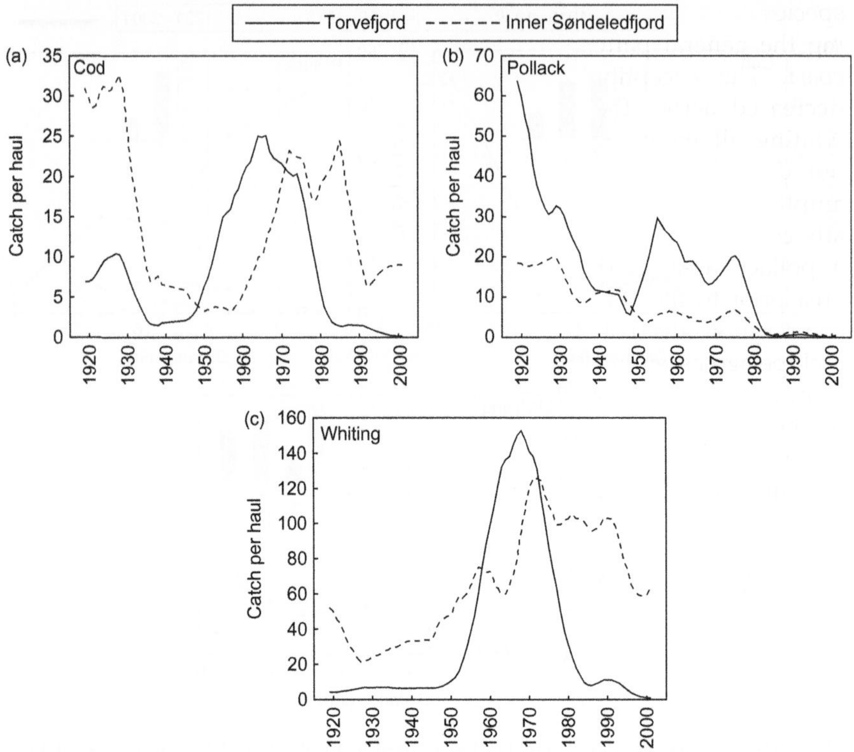

FIGURE 2.6 Smoothed (two times seven-year moving average) catches of (a) cod, (b) pollack and (c) whiting in Torvefjord (solid line) and in inner Søndeledfjord (hatched line) 1919–2001. To make the panels more readable, only the smoothed curves are presented.

reductions of cod and whiting from the second to the last period, whereas at location 95 the catches of both cod and whiting more than doubled.

2.3.2.3 Differences Between an Exposed and an Semi-enclosed Subarea

Effects on temporal abundance variation in relation to degree of exposure to coastal conditions are illustrated for the Torvefjord as a coastal archipelago regarded as exposed and the semi-enclosed inner Søndeledfjord (Fig. 2.2). In agreement with the pattern in the inner fjord-to-coast comparison shown in the previous section, pollack was the most abundant species in the exposed Torvefjord in the 1920s, and cod was most abundant in the enclosed inner Søndeledfjord. Both species decreased in the 1930s and increased again in the 1950s, following the general pattern in bottom flora coverage along the Skagerrak coast. The exception was the inner Søndeledfjord, where pollack decreased across the entire period. In Torvefjord, cod, pollack, and whiting all increased to relatively high levels in the 1960s to 1970s but decreased drastically again in the late 1970s. Except for larger amplitudes, the temporal patterns of abundance in Torvefjord generally concurred with those of the Skagerrak coast. However, from being a pollack area in the 1920s, Torvefjord changed to a cod–whiting area prior to the severe decline in the 1970s. Thus, in terms of species composition, the Torvefjord changed from a site with a typical coastal archipelago assemblage to a more fjord-like assemblage.

In inner Søndeledfjord the abundance of cod and whiting generally followed the same temporal development as observed in Torvefjord but with a delay of about 15 years and substantially smaller amplitudes of abundance variation.

2.3.3 Areas with Recruitment Collapses

In the three subareas Grenland, Holmestrandfjord, and the Inner Oslofjord, particularly strong and persistent declines in the abundance of 0-group gadoids occurred. These declines exceeded 90% of average abundances in preceding years of the time series and may be regarded as recruitment collapses.

2.3.3.1 *Grenland*

In Grenland, the beach seine sampling program started in 1953. In the mid-1960s the recruitment of 0-group cod, pollack, and whiting suddenly collapsed (Fig. 2.7). The average catches of the three species dropped by 92%, 97%, and 94%, respectively. Significant shifts ($p \leq 0.010$, STARS) occurred in 1966, 1965, and 1965 in cod, pollack, and whiting, respectively. In the years after the shifts, the recruitment varied without a trend at a very reduced level ($p \geq 0.295$, linear regression).

Reliable observations of bottom flora coverage in Grenland were often prevented by poor visibility. However, some good observations exist, and combined with descriptions of plants retained in the seine, a picture of changes in bottom flora coverage has emerged. The records observe that until the late 1950s relatively low bottom vegetation coverage consisting of eelgrass and brown macro-algae (approximately 15%) was recorded in this subarea. Gradually, the green alga *Monostromum oxyspermum* increased, and by the mid-1960s the coverage was so dense that the seine sometimes could not be recovered and

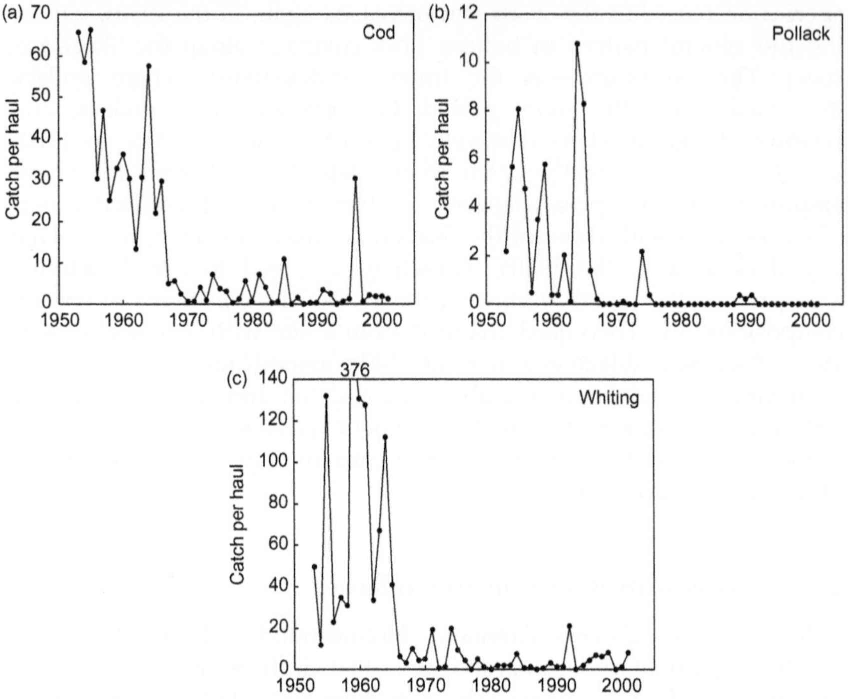

FIGURE 2.7 Average catches of (a) cod, (b) pollack, and (c) whiting at seven beach seine locations in Grenland, 1953–2001.

operated according with established procedures. Subsequently, the abundance of *M. oxyspermum* gradually decreased to about the same level as in the 1950s, and eelgrass and brown macro-algae have regained their former levels. *M. oxyspermum* forms a carpetlike cover on the bottom, whereas eelgrass forms a three-dimensional habitat in which juvenile fish can hide. Bottom flora coverage did not change abruptly as did 0-group fish abundance, and the collapse in fish recruitment was apparently decoupled from that of the bottom flora variation.

2.3.3.2 *Holmestrandfjord*

Sampling in Holmestrandfjord started in 1936. As in Grenland, the recruitment of the 0-group gadoids collapsed in the mid-1960s (Fig. 2.8) and remained at low levels up to the 2000s. Significant shifts ($p \leq 0.018$, STARS) were detected in 1967, 1965, and 1964 in cod, pollack, and whiting, respectively. After the shifts, there were no trends ($p \geq 0.225$). The bottom flora in Holmestrandfjord increased

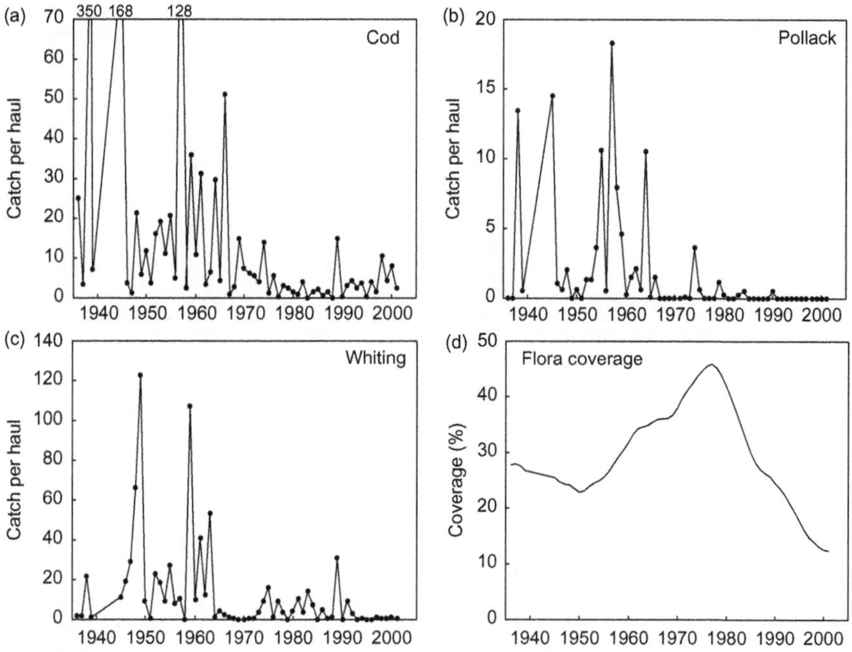

FIGURE 2.8 Average catches of (a) cod, (b) pollack, and (c) whiting at six beach seine locations in the Holmestrandfjord, 1936–2001, (d) the smoothed bottom vegetation coverage at the beach seine locations (two times seven-year moving average).

gradually until about 1980, followed by a gradual decrease. Hence, the recruitment collapses in the mid-1960 were decoupled from changes in bottom flora.

2.3.3.3 *Inner Oslofjord*

Ever since the sampling started in Inner Oslofjord in 1936, the abundance of 0-group gadoids (Fig. 2.9) has been at the same low level as in Grenland and Holmestrandfjord after the collapses described previously. An exceptionally strong year-class of cod occurred in 1938 and some relatively strong year-classes in other years, however. The indication of a very strong year-class of whiting in 1978 is due to a single exceptional catch of 1000 whiting at the outermost location close to the entrance of the Inner Oslofjord.

The commercial landings of cod in Inner Oslofjord dropped abruptly by approximately 85% in the early 1930s (Fig. 2.9d), suggesting that a similar collapse occurred in the Inner Oslofjord at that time (i.e., similar to those that occurred later in Grenland and Holmestrandfjord). STARS detected a shift in the landings in 1931 ($p < 0.001$). No bottom flora

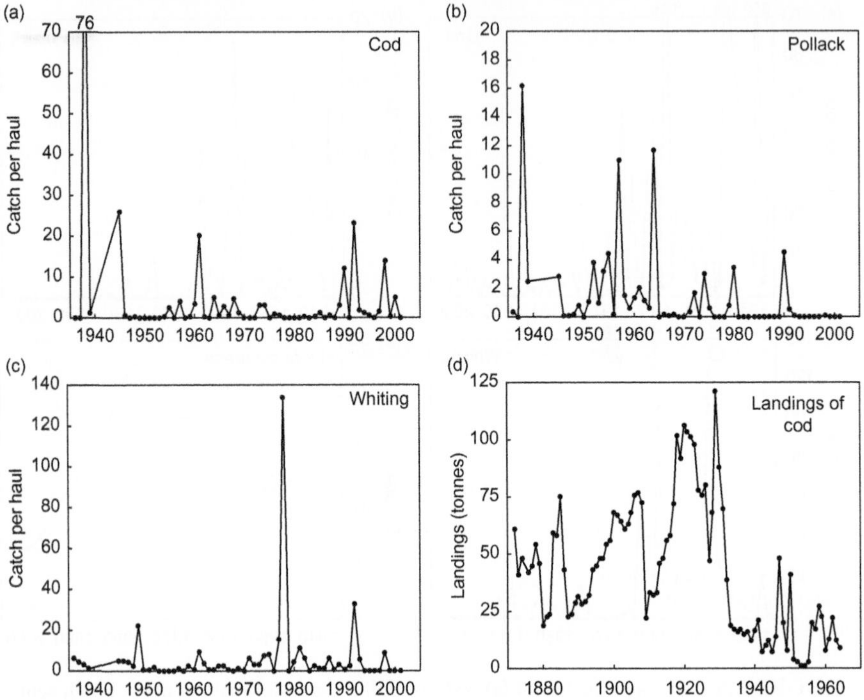

FIGURE 2.9 Average catches of (a) cod, (b) pollack, and (c) whiting at nine beach seine locations in Inner Oslofjord, 1936–2001, (d) commercial landings of cod from Inner Oslofjord, 1872–1964; (d) was modified from Ruud (1968b), with permission from Springer.

coverage data are presented here because no observations were recorded prior to the 1930s.

During the survey period from 1936 onwards, no significant shifts in the 0-group gadoid abundances were detected, nor were there trends ($p \geq 0.175$, linear regression) in cod and whiting. There was a decreasing trend in pollack ($p < 0.001$). However, the average catch dropped from merely 2.7 pollack per haul before 1965 (the last year with relatively high catches) to 0.4 after 1965.

2.3.4 Pooled Abundance of Gadoid and Non-Gadoid Fishes After the Recruitment Failures

In addition to there being no trends after the previously observed recruitment failures (except for the slight decrease in abundance in pollack in Inner Oslofjord), the pooled abundance of all the gadoids were

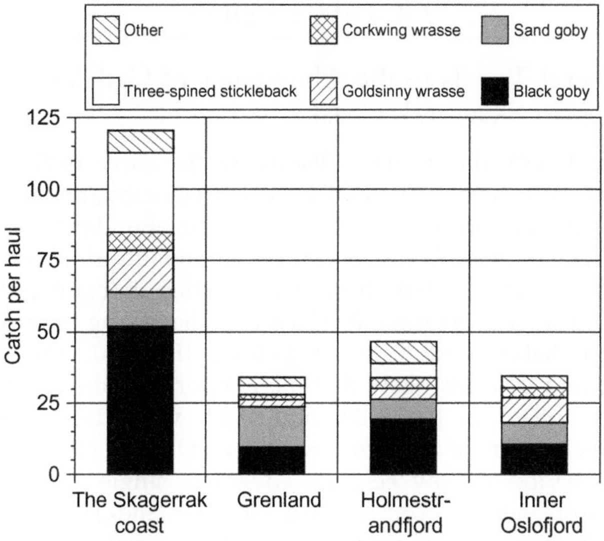

FIGURE 2.10 Catch of non-gadoid fishes in various areas along the Skagerrak coast 1989–2001 (black goby—*Gobius niger*, sand goby—*Pomatoschistus minutus*, goldsinny wrasse—*Ctenolabrus rupestris*, corkwing wrasse—*Symphodus melops*, three-spined stickleback—*Gasterosteus aculetus*).

about the same in all subareas ($p = 0.261$, ANOVA on log-transformed sum of cod, pollack, and whiting for individual beach seine hauls; 1 was added because of zero catches). The postcollapse average catches were 9.1, 8.4, and 11.6 gadoids per haul in Grenland, Holmestrandfjord, and Inner Oslofjord, respectively. In the relatively exposed coastal site Torvefjord, the average catch of gadoids after the decline in the late 1970s was 8.1 fish per haul (i.e., a very similar level to that in the locations where apparent collapses occurred).

The catches of non-gadoid fishes have been quantified systematically since 1989. Pelagic species that occur more sporadically in the sublittoral zone and small species that are not sampled reliably by the 15 mm meshes are not included in these analyses. The average catches of non-gadoid fishes between 1989 and 2001 (Fig. 2.10) were significantly higher along the Skagerrak coast than in the three subareas with severe recruitment failures ($p < 0.001$, ANOVA on log-transformed sum of all species per beach seine haul). However, there were no significant differences between the three areas with recruitment failures ($p \geq 0.556$, Fisher's PLSD). In Torvefjord, the average catches of non-gadoids between 1989 and 2001 were 27.5 fish per haul, similar to that in Grenland. Most of these fishes are small (<15 cm) and therefore potential prey of larger predator fishes, mainly gadoids.

2.4 DISCUSSION

2.4.1 General Trends in the Abundance of Gadoids Along the Norwegian Skagerrak Coast

Concurrent with the eelgrass disease in the early 1930s, the abundances of 0-group cod and pollack decreased substantially. With the recovery of the bottom flora in the 1950s, the abundances of cod and pollack increased again. Whiting was not abundant prior to and during the decrease in eelgrass but showed increasing abundance during the period of regrowth of eelgrass. Eelgrass beds have long been recognized as important habitats for 0-group gadoids (Blegvad, 1917; Gotceitas et al., 1997) and it seems likely that the temporary reduced abundances of cod and pollack in the 1930s and 1940s were linked to the pronounced decimation of bottom vegetation. On the other hand, even during the period of low eelgrass coverage, single years with high abundances of gadoids occurred (e.g., in 1938 and 1945 for both cod and pollack). Hence, while the impact of the vegetation changes seems the more likely explanation for the temporal pattern prior to the 1950s, other factors may also have contributed to low abundances in the 1930s and 1940s. In the mid-1970s though, both cod, pollack, and whiting showed marked reductions without parallel changes in the bottom vegetation.

Historically, 0-group cod and whiting were most abundant in fjords and inner coastal areas, while pollack, prior to the collapse in recent decades, was confined to the archipelago and outer fjords (Fig. 2.5). This is consistent with other reports (Fromentin et al., 1997, 1998; Hop et al., 1994). Until the mid-1970s pollack and whiting showed opposite temporal trends, and in the rather exposed Torvefjord there was a shift from high abundance of pollack in the 1920s to extraordinary high abundances of whiting and cod prior to the strong decline in all species in the mid-1970s (Fig. 2.6). This might have indicated interspecific competition as the mechanism behind the opposite trends in whiting and pollack until the mid-1970s, and also as a mechanism behind the distributional differences in the 0-group gadoids. However, in the subarea of Sandnesfjord and Risør archipelago the abundance of pollack was substantially reduced in the outer part of the fjord and the archipelago prior to the mid-1970s, without corresponding increases in the abundance of whiting and cod. The general increase in whiting abundance concurred with the recovery of eelgrass, yet whiting had very low abundances well before the decimation of eelgrass in 1933. This suggests that prior to the mid-1970s there was a temporal change in the recruitment conditions that were positive for whiting and negative for pollack.

The fisheries impact on recruitment levels and abundance of the gadoid species on the Skagerrak has not been monitored satisfactorily. Cod, pollack, and saithe, but not whiting and poor-cod, are target species of traditional small-scale hook and line and gillnet fisheries. A fisheries impact on the overall abundance of the target species is likely, but recruitment to the fisheries happens well after the 0-group stage; hence a direct exploitation effect on early juvenile survival would seem unlikely. Also, there is no major trawl or seine fishery along this coast that could have resulted in overfishing of 0-group juveniles of the target species nor on nontarget species such as whiting and poor-cod.

The relevant question in the context of this analysis, however, is whether the long-term patterns and in particular the observed abrupt declines in many areas could be a result of recruitment overfishing (i.e., reduced recruitment due to small spawning stocks). The abrupt decline in abundance of the nontarget species whiting observed along the entire coastline in the mid-1970s and the collapse in poor-cod are very unlikely to be attributed to recruitment overfishing. There is no evidence of any changes in fishing patterns or exploitation levels at that time. As there were no major changes in bottom flora in the mid-1970s either, the sudden drop in whiting suggests other abrupt ecological changes affecting recruitment. Although not equally conspicuous as in whiting, the abundance of cod decreases also in the mid-1970s, and in pollack the last year with relatively high abundance was in 1976. Coinciding with the declines in the gadoids, Johannessen and Dahl (1996a, 1996b) reported an abrupt decrease in bottom water oxygen along the entire Norwegian Skagerrak coast, which was attributed to structural changes in the pelagic community.

The observations suggest that the marked reduction of the gadoids along the Skagerrak coast in the mid-1970s (Fig. 2.4) was a regional phenomenon, but that the severe decline primarily occurred in the relatively open coastal waters. For example, in the enclosed inner Søndeledfjord, neither cod nor whiting decreased in abundance in the mid-1970s (Fig. 2.6). Because cod is the innermost species with respect to distribution (Fig. 2.5), this may partly explain why cod has the least clear-cut drop in abundance in the mid-1970s, while the recruitment of pollack, which tends to be distributed in open coastal waters, collapsed. In the relatively exposed Torvefjord, on the other hand, cod collapsed abruptly in the mid-1970s (Fig. 2.6a).

2.4.2 Areas with Recruitment Collapses

Three subareas (Grenland, Holmestrandfjord and Inner Oslofjord) experienced especially severe and abrupt declines in abundance of cod, whiting, and pollack (i.e., reductions to less than 10% of the preceding abundance level). Despite different timing, some characteristics were common between these subareas:

1. The collapses occurred abruptly.
2. Prior to the collapses the abundance of gadoids was generally high.
3. There were no obvious signs in any of the variables recorded, suggesting an imminent collapse.
4. The collapses occurred simultaneously for cod, pollack, and whiting (Grenland and Holmestrandfjord).
5. Although single years with good recruitment occurred, the abundances of 0-group gadoids remained at low levels after the collapses with no subsequent signs of recovery.
6. The abundance of other shallow-water fishes was significantly lower in these subareas than in adjacent areas sampled (Fig. 2.10).
7. After the collapses, the abundances of both gadoids and other shallow-water fishes were the same in all areas with recruitment collapses (no significant differences).

The abrupt nature of the collapses, the asynchronous timing between the three areas, and the observation that these collapses were local and not coastwide phenomena strongly suggest that exceptional and localized, and apparently lasting, changes in the survival conditions occurred. Overfishing is unlikely because no such abrupt exploitation changes occurred. Also, whiting collapsed too despite being virtually unexploited. In addition, the abundances of other nontarget littoral fishes were also consistently low in these areas (Fig. 2.10), despite the fact that the predation pressure on these relatively small fishes probably had decreased substantially due to the recruitment failure in the gadoids. Juvenile and adult gadoids are probably the dominating larger predator fishes in these waters as they constituted 80% of piscivorous fishes greater than 20 cm in the beach seine catches between 1919 and 2001.

The Inner Oslofjord is a special case in the sense that the beach seine sampling started in 1936 after the abrupt drop in the landings of cod. It is reasonable to conclude that there was a recruitment collapse around 1930 and that recruitment failure caused the decline in the fishery. The situation in Inner Oslofjord shares all the characteristics of Grenland and Holmestrandfjord after the collapses, that is the same low levels of 0-group gadoids, and of noncommercial fishes. Sporadic strong year-classes of cod do suggest that there is a potential for good recruitment (e.g., in 1938 and 1945). Ruud (1968a) found no changes in fishing patterns in Inner Oslofjord that could explain the abrupt decrease in landings of cod in the early 1930s and concluded that the decline in catch rates and abundance probably resulted from pollution.

Indeed, all three areas that experienced lasting recruitment failure were significantly affected by pollution (i.e., anthropogenically generated chemical contamination of various kinds). Concerns regarding pollution effects in Inner Oslofjord were expressed as early as the 1930s (see Dale et al. 1999; Ruud, 1968a). Considerable effort has been aimed at assessing the impact of pollution, including chemical contaminants and heavy nutrient inputs, on the marine environments of both Inner Oslofjord and Grenland. Although much of this work has been published in the gray literature, workshop results for both areas have been published in international journals (Bayne et al, 1988; Ruud, 1968b). Less work has been done in Holmestrandfjord, but some papers were published in international journals (e.g., Alve, 1995; Alve and Nagy, 1986). In all three areas the collapses in gadoid recruitment coincided with increasing pollution and, in particular, increased nutrient loading.

Increased nutrient loads may result in a shift from slow-growing sea grasses and large macroalgae to fast-growing macroalgae and ultimately to the dominance of phytoplankton (Duarte, 1995). With the exception of the decrease in bottom flora in the 1930s and 1940s, which was a result of an eelgrass disease (Short et al., 1988), no concurrent changes in bottom flora concomitant with the recruitment collapses in either Grenland or Holmestrandfjord were recorded. Turbidity may affect the foraging success of visual predators such as the gadoids (e. g., Utne, 1997) but also the predation risk of fish larvae (e.g., Fiksen et al., 2002). In Grenland the recruitment collapse occurred during a period with high turbidity. However, subsequently, as a response to reduced local nutrient loads, the visibility of the surface water increased substantially above the level at which the collapse occurred and also above the level when the gadoids were highly abundant (Johannessen and Dahl, 1996a). Alternatively, increasing incidence of anoxia may result from eutrophication (Diaz and Rosenberg, 2008). However, oxygen conditions were generally good in both Grenland (Johannessen and Dahl, 1996a) and Holmestrandfjord (author's unpublished data), that is, well above the 40% saturation threshold where cod and whiting have been found to migrate to escape low-oxygen water (Baden et al., 1990). Moreover, relatively good recruitment was observed in inner Søndeledfjord (Fig. 4.6) where close to anoxic conditions have prevailed below 20–30 m depth (Johannessen and Dahl, 1996a). These observations suggest that classical direct and most apparent effects of enhanced nutrient load do not explain the abrupt and persistent recruitment failures.

In the case of direct effects of heavy metals and organic contaminants on fish (including fish larvae, spawning products, and spawning behavior), different tolerance limits for the various species would be expected. Hence, simultaneous recruitment collapses in cod, pollack, and whiting suggest that the collapses are linked to major ecosystem changes rather than the direct impact of contaminants on the fish.

2.4.3 Eutrophication as a Probable Common Cause

The marked decrease in recruitment of gadoids along the Skagerrak coast in the mid-1970s followed the same pattern as in the three polluted areas discussed previously, with similarly abrupt, simultaneous decreases in cod, pollack, and whiting. However, there were differences between areas, from moderate changes in the enclosed inner Søndeledfjord to a collapse in the more-exposed Torvefjord. Interestingly, the habitat in Torvefjord changed from a pollack habitat to a cod and whiting habitat prior to the collapse. In Grenland and Holmestrandfjord too, cod and whiting dominated prior to the collapse.

The similarity between the areas and coastwide observations suggests that a common mechanism underlies the recruitment declines and collapses that happened early in the most polluted areas and only in the 1970s in less-polluted coastal sites. Concomitant with the marked decrease in recruitment of gadoids along the Skagerrak coast in the mid-1970s, an equally abrupt and persistent reduction in bottom water oxygen content was observed along the entire coast, which was attributed to the ongoing eutrophication in these waters (Johannessen and Dahl, 1996a, 1996b). Aure et al. (1996) estimated the oxygen consumption in one of the fjords to have increased by 50% at that time. The abrupt change in bottom water oxygen has been proposed as a signal of changes in the planktonic community structure that has resulted in increased sedimentation of phytoplankton and phytodetritus (Johannessen and Dahl, 1996a).

Other types of pollutants often accompany eutrophication, as in the highly industrialized Grenland area (Bryne et al., 1988). However, studies of dated sediments from Inner Oslofjord revealed that concentrations of heavy metals and more modern contaminants like PCB and DDT were low at the time of the collapse around 1930, but subsequently increased to relatively high concentrations (Konieczny, 1994). Already in 1917, however, it was observed that the phytoplankton population outside the harbor of Oslo was unusually large, and it was suggested that this might be due to the discharge of sewage from the city into this semi-enclosed fjord (reviewed by Ruud, 1968a). In Oslo the number of water closets increased from 4,788 in 1916 to 48,937 in 1936 (Braarud, 1945). In the early 1930s it became evident that eutrophication was an important factor for the biology of the whole Inner Oslofjord, and a pronounced effect on the phytoplankton of the fjord was described (Ruud, 1968a). This is strong evidence suggesting that the recruitment collapse in Inner Oslofjord around 1930 was somehow related to effects of eutrophication. In Grenland too, the recruitment collapse occurred during a period of substantially increasing nutrient loading (Johannessen and Dahl, 1996a). Holmestrandfjord (Sandebukta) is considered to be a polluted area due to local input of nutrients and organic matter that have resulted in significant changes in the foraminifera fauna during the past 100 years (Alve, 1995; Alve and Nagy, 1986).

The marked reductions in gadoid recruitment in the mid-1970s along the Skagerrak coast varied locally from limited effects in the semi-enclosed Søndeledfjord to severe recruitment failures in the more exposed Torvefjord. There are no major industries that could potentially have contaminated Torvefjord. Eutrophication and its effects, on the other hand, are regional phenomena resulting from both local sources of nutrients (mainly from freshwater input) and long-distance transport of nutrients from the southern North Sea, Kattegat, and the Baltic (Johannessen and Dahl, 1996a). Hence, not only the most heavily polluted subareas but all areas with recruitment collapses and declines were probably subjected to substantial and increasing eutrophication at the time of the major abundance declines. Enhanced nutrient loading is the only known factor common to the various subareas. Therefore, from the present evidence I conclude that ecological mechanisms related to or caused by eutrophication are a primary cause of the observed recruitment failures (at least failure events not related to the probable negative impact of the eelgrass disease in the 1930s).

In summary, a number of candidate causes of the recruitment collapses are unlikely, such as overfishing, hypoxia, habitat degradation through changes in bottom vegetation, and direct impact of contaminants on fish and spawning products. Based on the strong similarities between the incidents and the fact that eutrophication was the only factor common to the various subareas, gradually increasing nutrient loads is put forward as a more likely agent somehow leading to the recruitment failures. As nutrients have a direct influence on phytoplankton and zooplankton community structure and production on which the planktivorous juvenile gadoids depend, bottom-up ecological relationships, and community dynamics and processes should be explored to investigate further mechanisms underlying recruitment variation, abrupt shifts, and persistence of apparently low survival rates

The mechanism underlying the recruitment collapses is subjected to comprehensive testing in Chapters 3 and 4, and Chapter 5 reviews direct and indirect evidence of abrupt changes in the plankton community.

References

Alve, E., 1995. Benthic foraminiferal responses to estuarine pollution: a review. J. Foraminiferal Res. 25, 190–203.

Alve, E., Nagy, J., 1986. Estuarine foraminiferal distribution in Sandebukta, a branch of the Oslo Fjord. J. Foraminiferal Res. 21, 261–284.

Aure, J., Danielssen, D., Svendsen, E., 1998. The origin of Skagerrak coastal water off Arendal in relation to variations in nutrient concentrations. ICES J. Mar. Sci. 55, 610–619.

Aure, J., Danielssen, D.S., Sætre, R., 1996. Assessment of eutrophication in Skagerrak coastal waters using oxygen consumption in fjordic basins. ICES J. Mar. Sci. 53, 589–595.

Baden, S.P., Loo, L.-O., Pihl, L., Rosenberg, R., 1990. Effects of eutrophication on benthic communities including fish: Swedish west coast. Ambio 3, 113–122.

Bayne, B.L., Clarke, K.R., Gray, J.S., 1988. Biological effects of pollutants. Result from a practical workshop. Mar. Ecol. Prog. Ser. 46, 265–278.

Blegvad, H., 1917. Om fiskenes føde i de danske farvande innden for Skagen. Beretning til Landbrugsministeriet 24, 17–72 (in Danish).

Braarud, T., 1945. A phytoplankton survey of the polluted waters of inner Oslo Fjord. Hvalråd. Skr. Sci. Results Mar. Biol. Res. 28, 1–142.

Caddy, J.F., 1993. Toward a comparative evaluation of human impacts on fishery ecosystems of enclosed and semi-enclosed seas. Rev. Fish. Sci. 1, 57–95.

Dale, B., Thorsen, T.A., Fjellså, A., 1999. Dinoflagellate cysts as indicators of cultural eutrophication in the Oslofjord, Norway. Est. Coast. Shelf Sci. 48, 371–382.

Diaz, R.J., Rosenberg, R., 2008. Spreading dead zones and consequences for marine ecosystems. Science 321, 926–929.

Duarte, C.M., 1995. Submerged aquatic vegetation in relation to different nutrient regimes. Ophelia 41, 87–112.

Fiksen, Ø., Aksnes, D.L., Flyum, M.H., Giske, J., 2002. The influence of turbidity on growth and survival of fish larvae: a numerical analysis. Hydrobiologia 484, 49–59.

Fromentin, J.M., Stenseth, N.C., Gjøsæter, J., Bjørnstad, O.N., Falck, W., Johannessen, T., 1997. Spatial patterns of the temporal dynamics of three gadoid species along the Norwegian Skagerrak coast. Mar. Ecol. Prog. Ser. 155, 209–222.

Fromentin, J.M., Stenseth, N.C., Gjosæter, J., Johannessen, T., Planque, B., 1998. Long-term fluctuations in cod and pollack along the Norwegian Skagerrak coast. Mar. Ecol. Prog. Ser. 162, 265–278.

Gjøsæter, J., Danielssen, D.S., 1990. Recruitment of cod (*Gadus morhua*), whiting (*Merlangius merlangus*) and pollack (*Pollchius pollchius*) in the Risør area on the Norwegian Skagerrak coast 1945 to 1985. Flødevigen Rapportser. 1, 11–31.

Gotceitas, V., Fraser, S., Brown, J.A., 1997. Use of eelgrass beds (*Zostera marina*) by Juvenile Atlantic cod (*Gadus morhua*). Can. J. Fish. Aquat. Sci. 54, 1306–1319.

Hop, H., Gjøsæter, J., Danielssen, D.S., 1994. Dietary composition of sympatric juvenile cod, *Gadus morhua* L., and juvenile whiting, *Merlangius merlangus* L., in a fjord of southern Norway. Aquac. Fish. Manage. 25, 49–64.

Islam, M.S., Tanaka, M., 2004. Impacts of pollution on coastal and marine ecosystems including coastal and marine fisheries and approach for management: a review and synthesis. Mar. Pollut. Bull. 48, 624–649.

Johannessen, T., Dahl, E., 1996a. Declines in oxygen concentrations along the Norwegian Skagerrak coast 1927–1993: A signal of ecosystem changes due to eutrophication? Limnol. Oceanogr. 41, 766–778.

Johannessen, T., Dahl, E., 1996b. Historical changes in oxygen concentrations along the Norwegian Skagerrak coast: A reply to the comment by Gray and Abdullah. Limnol. Oceanogr. 41, 1847–1852.

Johannessen, T., Dahl, E., Falkenhaug, T., Naustvoll, L.J., 2012. Concurrent recruitment failure in gadoids and changes in the plankton community along the Norwegian Skagerrak coast after 2002. ICES J. Mar. Sci. 69, 795–801.

Johannessen, T., Sollie, A., 1994. Overvåking av gruntvannsfauna på Skagerrakkysten–historiske svingninger i fiskefauna 1919–1993, og ettervirkninger av den giftige algeoppblomstringen i mai 1988. Fisken og Havet 10, 1-91 (in Norwegian, <http://brage.bibsys.no/imr/handle/URN:NBN:no-bibsys_brage_3791)>.

Konieczny, R.M., 1994. Miljøgiftundersøkelser i Indre Oslofjord. Delrapport 4. Miljøgifter i sedimenter. NIVA overvåkningsrapport 561/94 (82-577-2564-1), 1-134 (in Norwegian).

Methven, D.A., Schneider, D.C., 1998. Gear-independent patterns of variation in catch of juvenile Atlantic cod (Gadus morhua) in costal habitats. Can. J. Fish. Aquat. Sci. 55, 1430–1442.

Riley, J.D., Parnell, W.G., 1984. The distribution of young cod. Flødevigen Rapportser. 1, 563–580.

Rodionov, S., Overland, J., 2005. Application of a sequential regime shift detection method to the Bering Sea ecosystem. ICES J. Mar. Sci. 62, 328–332.

Ruud, J.T., 1968a. Introduction to studies of pollution in the Oslofjord. Helgol. Wiss. Meeresunters. 17, 455–461.

Ruud, J.T., 1968b. Changes since the turn of the century in the fish fauna and fisheries in the Oslofjord. Helgol. Wiss. Meeresunters. 17, 510–517.

Short, F.T., Ibelings, B.W., den Hartog, C., 1988. Comparison of a current eelgrass disease to the wasting disease in the 1930s. Aquat. Bot. 30, 295–304.

Tveite, S., 1971. Fluctuations in year-class strength of cod and pollack in southeastern Norwegian coastal waters during 1920–1969 Fiskeridir. Skr. Ser. Havunders. 16, 65–76.

Tveite, S., 1992. Prediction of year-class strength of coastal cod (*Gadus morhua*) from beach seine catches of 0-group. Flødevigen Rapportser. 1 (1992), 17–23.

Utne, A., 1997. The effect of turbidity and illumination on the reaction distance and search time of the marine planktivore *Gobiusculus flavescens*. J. Fish Biol. 50, 926–938.

Williams, C., 1996. Combatting marine pollution from land-based activities: Australian initiatives. Ocean Coast. Manage. 33, 87–112.

CHAPTER

3

Causes of Variation in Abundance, Growth, and Mortality in 0-Group Gadoids After Settlement and a Hypothesis Underlying Recruitment Variability in Atlantic Cod

3.1 INTRODUCTION

3.1.1 Background

In Chapter 2, results from the extensive time series of annual beach seine monitoring program run since 1919 along the Norwegian Skagerrak coast were presented. Several incidents of abrupt and persistent recruitment collapses in gadoids were documented in relation to increasing eutrophication. The collapses occurred simultaneously in the gadoids, and after the collapses there have been no signs of recovery nor have other littoral or sublittoral fishes replaced the gadoids. A number of proposed underlying causes of the observed declines were considered unlikely, such as, overfishing, hypoxia, habitat degradation through changes in bottom vegetation, and direct impact of contaminants on fish and spawning products. Based on the strong similarities between the incidents in different localities and the observation that eutrophication was the only factor common to the various sites, it was concluded that the recruitment collapses were somehow related to

From an Antagonistic to a Synergistic Predator Prey Perspective.
DOI: http://dx.doi.org/10.1016/B978-0-12-417016-2.00003-8

gradually increasing nutrient loads and the influence of enhanced nutrient levels on pelagic food webs in which gadoids are members. The gadoids depend on pelagic prey during early life stages, and it was suggested that increasing nutrient loads impacted phyto- and zooplankton communities in a manner that abruptly and then persistently deprived the gadoids of their natural prey.

3.1.2 Mechanisms Underlying Recruitment Variability

Ever since the pioneering work of fisheries biologists around the turn of the twentieth century revealed substantial inter-annual variability in year-class strength of a number of fish species (Hjort, 1914), explaining recruitment variability has been a major challenge in fisheries research and marine ecology. Recruitment variability is one of the dominant causes of abundance fluctuations in marine fishes. Early laboratory experiments had shown that the time of complete yolk absorption is a critical phase for survival of fish larvae (Fabre-Domergue and Bietrix, 1897). By combining this information with field observations on the spawning grounds of Arcto-Norwegain cod stock (*Gadus morhua*), Hjort (1914) suggested that a mechanism regulating year-class strength could be food availability, with a crucial factor being the degree of match/mismatch between larval abundance and food availability, the so-called critical period hypothesis (May, 1974). Despite more than 100 years of research, the mechanisms underlying recruitment variability remain largely unknown (Anderson, 1988; Kendall and Duker, 1998). There is, however, general consensus that most of recruitment variability occurs at an early stage in the life cycle of fish, and the critical period hypothesis (or modifications of this hypothesis) remains the most generally accepted explanation. Most recruitment studies have consequently focused on the larval phase (Blaxter, 1974; Blaxter et al., 1989; Lasker and Sherman, 1981). Estimating survival or mortality rates of larval and post-larval stages of wild marine fish at realistic temporal and spatial scales is, however, very difficult. Hence, there are relatively few direct studies of inter-annual variability in the mortality rates of wild fishes during these early life stages.

3.1.3 Settlement, Growth, and Recruitment

To disclose the mechanism underlying the recruitment collapses in the gadoids, the mechanism that generates recruitment variability has to be resolved as well. Atlantic cod was chosen as model species for these studies because it was the only gadoid that settled in sufficient numbers and could be followed over several months after settlement.

Based on the beach seine survey data, abundance indices of 0-group gadoids were obtained annually since 1919 along the Norwegian

Skagerrak coast, and positive correlations between abundance of 0-group in September and I-group the subsequent year suggests that year-class abundance has mainly been determined by that time (see Chapter 2). However, there have been no systematic studies of mortality and growth between the spring spawning season and the beach seine sampling period in the latter half of September. The abundance of recruits in September results from the production of progeny by the number of spawners taking part in the preceding spawning season and/or the rate of mortality between spawning time and September. Due to the inherent difficulty of obtaining reliable estimates of survival and mortality rates in fish larvae in the field, this study focuses on the subsequent period (i.e., from settlement onwards to September). This allows testing of whether the year-class strength is determined before or after settlement (e.g., high correlation between the number of settlers and abundance in September would indicate that the period prior to settlement is important). Furthermore, the hypothesis that size-specific growth and mortality rates interact to determine survivorship in fish (e.g., Shepard and Cushing, 1980; Ware, 1975) can be tested during the period from settlement to when the year-class strength is mainly determined in September. This hypothesis is related to the combined impact of predation and food supply affecting growth rate. Because food supply may be affected by competition, the abundance of potential competitors of cod has been described. The abundance for these fishes, which are of the same size as settled cod, may also be alternative prey for larger predators and thus have an impact on the predation risk in cod. Furthermore, settlement in other gadoids is also described because this may shed light on the mechanisms governing the historical changes in gadoid abundances.

3.2 METHODS

The analyses presented in this chapter are based on independent beach seine investigations performed recently, as well as the data from the historical annual beach seine survey (Chapter 2). Based on analyses of the historical beach seine data and direct observations of fish behavior during the sampling process, it was concluded that the beach seine sampling is an adequate method for obtaining indices of fish abundances in shallow water (Chapter 2).

3.2.1 Settlement and Growth

The recent data were collected in the period 1997–1999 at six locations near the town of Arendal in southeast Norway (Fig. 3.1). The aim was

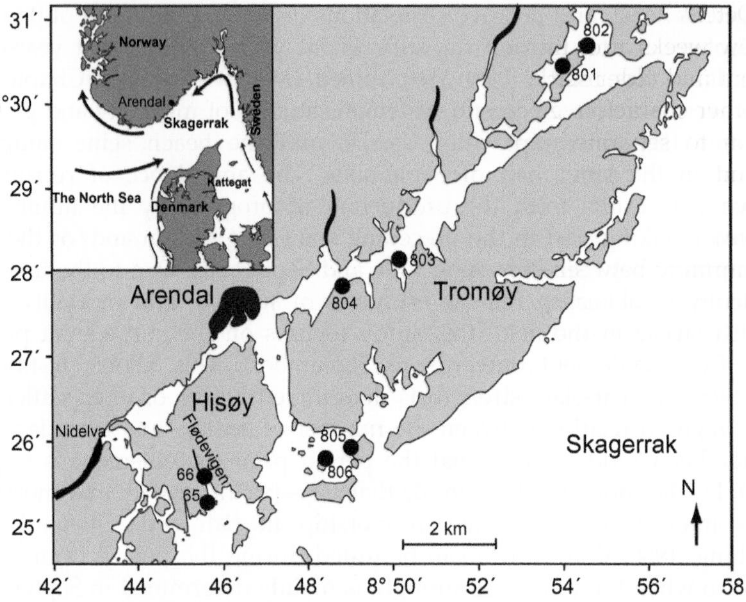

FIGURE 3.1 Beach seine sampling locations near Arendal. Locations 801–806 were sampled during this three-year study, and locations 65 and 66 were sampled during the annual beach seine survey in September.

to study settlement and growth in gadoids and the abundance of other littoral fishes and invertebrates. All sites sampled had a dense cover of eelgrass (*Zostera marina*), an important habitat for 0-group gadoids and other littoral fishes (Blegvad, 1917; Gotceitas et al., 1997). To sample the smallest life stages of the gadoids occurring near the seabed, a finer mesh beach seine than the traditional seine used during the regular annual survey was used (i.e., stretched mesh widths were 10 mm rather than 15 mm). In all other respects the two seines were identical (see Chapter 2).

In 1997, weekly sampling was carried out from late May to mid-August, followed by biweekly sampling until the end of September. The annual beach seine survey is carried out in the latter half of September. In 1998 and 1999, biweekly sampling was carried out from mid-April until the end of September. To evaluate the potential impact of frequent sampling on the abundance estimates, all fish were removed from three of the locations, and the fish were kept alive and released again at the other three locations. This procedure was adopted in 1997 and 1998, yet with a change between years with respect to which locations had the fish removed. Because no significant differences were observed between the two sampling strategies, this practice was abandoned in 1999.

Determining the size at settlement was not straightforward, however. A few weeks after the onset of settlement, the samples consisted of both recent and older settlers that could not readily be distinguished by size or other characters. Although previous studies used changes in pigmentation to separate newly settled cod from older settled conspecifics (Grant and Brown, 1998a; Tupper and Boutilier, 1995), no empirical evidence was presented demonstrating that pigmentation changes are related to settlement and not to other factors such as size. In this study, the approximate size range at settlement was obtained by subjectively evaluating the length frequency distributions from the onset of settlement to well beyond peak abundance.

3.2.2 Growth and Year-Class Strength in 0-Group Cod

Sampling over a three-year period did not provide sufficient inter-annual data to establish a reliable relationship between growth and year-class strength nor the potential impact of inter-annual variation in predation rate on the recruitment of 0-group cod. Therefore, the historical data (1919–2001) were used to study these aspects.

Presently approximately 130 beach seine locations are included in the annual beach seine sampling program, of which 38 locations have been sampled since 1919 (except during World War 2, 1940–1944). The relationship between year-class strength and size in 0-group cod was described on the basis of these 38 locations, situated between Torvefjord west of Kristiansand and Kragerø (Fig. 2.2). These locations were the same as those previously used to describe general trends in the abundance of 0-group gadoids along the Norwegian Skagerrak coast (Chapter 2).

Abrupt and persistent recruitment collapses in the gadoids were observed in Inner Oslofjord around 1930 (Fig. 2.9), in the Grenlandfjords (Fig. 2.7), and Holmestrandfjord (Fig. 2.8) in the mid-1960s. A less-severe recruitment failure occurred along the entire Skagerrak coast in the mid-1970s (Fig. 2.4). The sizes of the 0-group gadoids before and after the recruitment failures were analyzed based on historical data between 1953 and 2001. The year 1953 was chosen as a reference because this was the first year of sampling in the Grenlandfjords. No data were available from the period prior to the much earlier collapse in Inner Oslofjord (approximately 1930).

3.2.3 Predation

The potential impact of predation on recruitment of 0-group cod was studied by correlating the abundance of the potential predators with the recruitment index of 0-group cod from the historical beach seine data (1919–2001). Piscivorous fishes are generally three to four times larger than their prey (Bogstad et al., 1994; Grant and Brown, 1998b). Between settlement and mid-August, the majority of 0-group cod ranged between 3 and 10 cm. Potential predators were other abundant and common fishes that are sufficiently large (≥ 20 cm in September). These are ≥I-group cod, pollack (*Pollachius pollachius*), and saithe (*Pollachius virens*), and possibly also larger individuals of 0-group saithe. Sea trout is only caught in low numbers (average 0.2 per haul) but is large enough to predate on 0–group cod and was therefore included in the analyses.

The abundance of ≥I-group gadoids is generally much lower than that of the 0-group (Table 3.1). Therefore, to obtain a more precise estimate of the abundance of potential predators, all beach seine locations sampled over periods in excess of 20 years were included (average 81 locations yr^{-1}, i.e., the same locations that were used to assess the quality of the beach seine sampling; Chapter 2). Catch per haul was used as an index of abundance of both 0-group cod and their potential predators. To stabilize variance, the year-class indices were log-transformed (e.g., Zar, 1974). Pearson correlation analysis was carried out on the original abundance data series as well as on the data series that had been de-trended using a third-order polynomial regression.

3.3 RESULTS

3.3.1 Settlement and Growth in the Gadoids

The majority of cod in the six Arendal sampling locations settled in May and June at lengths of 2.5–5 cm TL (Total Length; Fig. 3.2b). The abundance of newly settled cod peaked in June at approximately the same level and time in the three years, 1997–1999 (Fig. 3.2a). The curves were generally smooth during the settlement period, suggesting that the beach seine provided relatively good estimates of newly settled cod. The increased abundance in July 1997 was only observed at one of the locations and was probably not caused by new settlement, as the cod were ≥7 cm. The location is situated at the outer part of a shallow bay (1–2 m deep) with eelgrass, where the bottom slopes relatively steeply into deeper water. Increased abundance coincided with heating of the surface water to > 20° C, and was probably caused by cod from the bay seeking colder water along the slope.

The growth of cod was about the same in the three years (Fig. 3.2b), and average length in September fell within the normal range for cod near Arendal (i.e., average of all 0-group cod sampled at locations 65

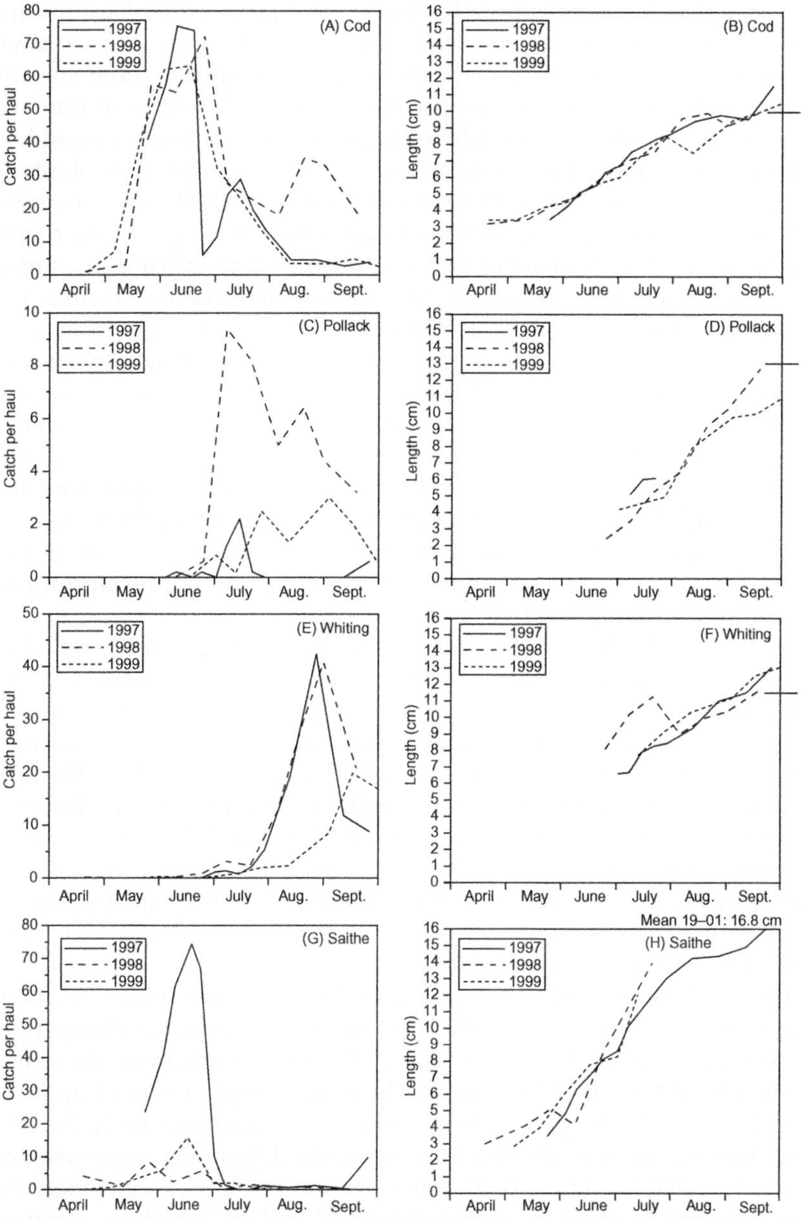

FIGURE 3.2 Average catch (left panels) and growth (right panels) in 0-group gadoids from May to September at six beach seine locations near Flødevigen 1997–1999. Horizontal lines at the right side of the length panels indicate average length about September at two beach seine locations near Flødevigen between 1919 and 2001. Average length of 0-group saithe is 16.8 cm. Notice the different scales.

and 66 in Flødevigen [Fig. 3.1] between 1919 and 2001). Despite that almost equal numbers settled at the same time of the year and had similar growth, the resultant year-classes were substantially different. The 1998 year-class apparently ended up much stronger than those in 1997 and 1999 ($p < 0.001$, ANOVA on individual catches after mid-August). This is consistent with the general inter-annual abundance differences observed along the Skagerrak coast during these years (Fig. 2.4a). The growth rate of cod as reflected in the population average length was about linear from early settlement to September and was estimated to 0.55 mm d^{-1} using simple linear regression on length data from all three years.

Young-of-the-year pollack appeared in the beach seine catches from the end of June (Fig. 3.2c), approximately two months later than cod. Settlement peaked in July at a length range of approximately 2.5–5 cm TL (Fig. 3.2d). Only the 1998 year-class was sufficiently abundant to determine growth rate. Growth was rapid (1.25 mm d^{-1} in 1998), and by September the length was similar to the average at locations 65 and 66 between 1919 and 2001. Although the year-class strengths were poor in all three years, the abundances of the year-classes 1997–1999 was about the same in the Arendal sites as along the entire Norwegian Skagerrak coast (Fig. 2.4b).

In 1997 and 1998 whiting (*Merlangius merlangus*) mainly settled in August at lengths of about 6–11 cm TL (Figs. 3.2e, f; maximum length uncertain). In 1998 there was a marked drop in the length of whiting in August, probably reflecting limited settlement during June and July, followed by a major settlement event in August. In 1999, settlement occurred about one month later at lengths of about 8–13 cm. The average growth rate of the population was 0.72 mm d^{-1} (length data from all years, except June–July 1998). The abundances of the various year-classes measured in September at the Arendal sites were in close agreement with that measured along the Skagerrak coast during the three years 1997–1999 (Fig. 2.4c).

The beach seine catches of 0-group saithe are usually minor, and during the period about 1935–1970 the species was virtually absent from the waters sampled (Fig. 2.4d). The 1997–1999 observations show that saithe settled from mid-May to mid-June, at a length range of approximate 2.5–8 cm TL (Figs. 3.2g, h; maximum size uncertain). Only the 1997 year-class was sufficiently abundant to permit description of growth and showed a relatively fast growth rate of 1.05 mm d^{-1}. By September, this year-class had reached an average length of 16.2 cm, which by comparison with the historical data appears normal for the subarea sampled.

Settlement of poor-cod (*Trisopterus minutus*) at the six locations near Flødevigen was insufficient to allow detailed studies. However, by combining beach seine data sampled between 1993 and 1999 at various locations along the Norwegian Skagerrak coast, it was possible to describe settlement for this species too. In this period a total of 113 specimens of 0-group poor-cod were sampled in the 192 beach seine hauls carried out in June, 248 hauls in July, 145 hauls in August, and 356 hauls in September. Of these 113 specimens, none occurred in June, 2 in July, 84 in August, and 27 specimens in September. Hence, poor-cod seems mainly to settle from August onwards. In mid-August the mean length of poor-cod was 4.5 cm (range 2.4–6.0 cm), and by mid-September the mean length had increased to 6.9 cm (5.0–8.8), indicating a growth rate of approximately 0.8 mm d^{-1}.

In conclusion, the growth rate was high in pollack and saithe, intermediate in whiting (and poor-cod), and slow in cod. The average weights of 0-group cod, pollack, whiting, and saithe from locations 65 and 66 in September for the period 1919–2001 were 7.8 g, 13.9 g, 16.6 g, and 46.6 g, respectively (estimated weight of the average length fish from author's unpublished length-weight relationship data). The facts that cod and saithe settle at about the same time and same size and that saithe is about eight times heavier than cod by September emphasize the slow growth in 0-group cod in these waters.

3.3.2 Seasonal Patterns in the Abundance of Littoral Fishes and Invertebrates

The dynamics of the numerically most important littoral fishes and invertebrates sampled by the beach seine (i.e., potential prey of cod and other fishes and/or competitors with cod) are illustrated for the period June–September 1997–1999 in Fig. 3.3. In all three years, marked reductions in the overall abundance of fish and invertebrates were observed from June to July, followed by substantial increases in August. This temporal pattern is mainly linked to the occurrence of annual gobies (*Gobidae*) that spawn in summer and attain adult sizes of 35–60 mm. The overall abundance of fish and invertebrates in June and July was highest in 1998, with a maximum abundance of more than 5000 individuals per beach seine haul, whereas in the autumn the abundance of fish and invertebrates (new generation) were substantially higher in 1999 and 1997 compared with 1998, with a maximum abundance of approximately 47,000 individuals per haul in September 1999.

Transparent goby (*Aphya minuta*), which is pelagic (Wheeler, 1969), was dominating during early summer when the I-group (average approximately 4.5 cm, TL) probably gathers in shallow water to spawn Shortly after spawning, the adults die, a feature that appears to be common among annual gobies (Mesa et al., 2005, and references therein). The abundance in June varied between approximately 2000 in 1997 and 4500 in 1998. The new generation of transparent goby is not sampled in high numbers in littoral and sublittoral waters later in the summer.

The second most abundant species in June was the I-group two-spot goby (*Gobiusculus flavenscens*, mean size approximately 4.0 cm TL on 1 July), which varied from about 300 per haul in 1997 to 100 in 1999 in the latter half of June. The new generation appeared in August and peaked in the first half of September every year, varying from approximately 9,000 per haul in 1998 to 42,000 in 1999. Mortality from peaks of abundance in autumn to spring appears to be high; the abundance was reduced from 42,000 0-group in 1999 to 13 I-group in June 2000 (additional sampling on 6 June 2000). It should be noted that the smallest life stages of 0-group two-spot and other gobies were not retained by even the 10 mm mesh seine. They were herded in front of the slow-moving seine and were observed escaping through the meshes just before the final recovery of the seine.

Sand goby (*Pomatoschistus minutus*) is a biannual species (Hesthagen, 1977), but only a few individuals survive to age 2 (Healey, 1971). In agreement with this, the abundance pattern of this species reflected the occurrence of I-group specimens in June, followed by a new generation appearing July–August. The abundance in the latter half of June of specimens of average length of 5.1 cm TL varied from 17 per haul in 1999 to 77 in 1997. Young-of-the-year sand goby dominated the catches in the first part of July in 1997 (approximately 85% by numbers), whereas in 1998 the new generation started to appear about one month later. Maximum abundance was observed in August.

The abundance of I-group painted goby (*Pomastoschistus pictus*) varied between 27 per haul in 1999 and 60 in 1998 in the latter half of June, and average length was approximately 3.8 cm TL. Young-of-the-year started to appear in August and peaked in August/September with abundances varying from about 400 in 1997 to 2000 in 1999.

Three-spined stickleback (*Gasterosteus aculetus*) spawn in late spring and early summer (Sokolowska and Kulczykowska, 2006), and the majority of the fish die after spawning at age 1 (Mori and Nagoshi, 1987). Accordingly, sticklebacks with an average length of approximately 5.6 cm TL (range 4.6–7.1 cm) occurred in relatively low numbers of 2–44 specimens/haul in June. In 1997 the new generation started to appear in July, and the average length was 3.0 cm TL (range 1.9–4.1 cm), and in 1998 about one month later. Peak abundance occurred in August, varying from about 700 per haul in 1999 to 3300 in 1997.

Common prawn (*Palaemon serratus*) totally dominated the decapod crustacean samples. In addition, a few *Palaemon elegans* and occasionally brown shrimps (*Crangon crangon*) were recorded. The abundance of common prawn varied between 5–78 specimens per haul in June, but increased from July onwards. Peak abundance occurred in September, varying from about 400 per haul in 1998 to 12,000 in 1997.

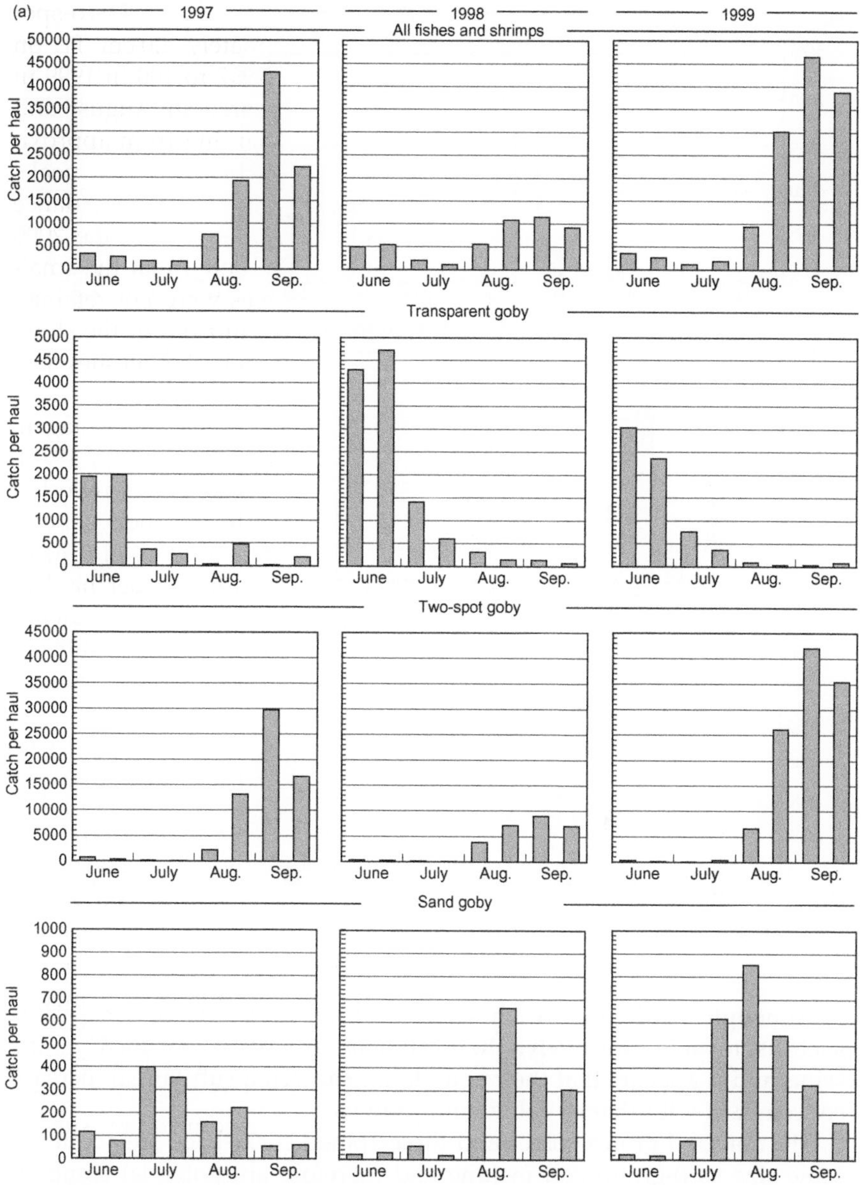

FIGURE 3.3 Average half-monthly catches of the most numerous non-gadoid species caught in beach seine at six locations near Flødevigen in June to September 1997–1999.

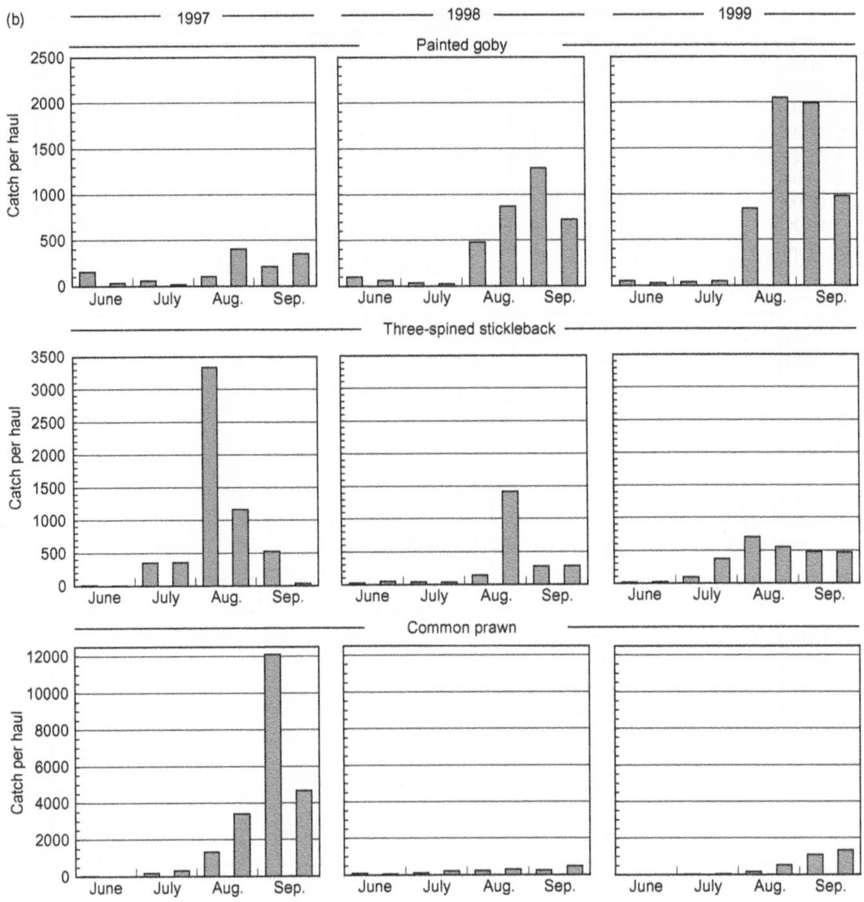

FIGURE 3.3 *(Continued.)*

Although the 10 mm beach seine does not catch smaller organisms such as mysids effectively, these animals are partly retained in the seine, and it appears that their abundance showed a substantial increase concomitant with that of gobies and prawns.

Most of the I-group gobies and three-spined stickleback are about the same size as 0-group cod in June and therefore are potential competitors but also alternative prey for larger predators. During peak abundance in June, cod constituted 2.8%, 1.4%, and 2.1% of the number of similar-sized fish in 1997, 1998, and 1999, respectively, and in the first part of July 2.2%, 3.1%, and 2.4%. Although the gadoids are probably the most abundant spring-spawners among the littoral fishes in the waters sampled, they are insignificant both in terms of numbers and biomass when compared with co-occurring summer-spawners.

3.3.3 Predation

In the period 1997–1999, the catches at the six Arendal locations of assumed potential predators on newly settled 0-group cod (i.e., ≥ I-group gadoids and sea trout) were only ≤ 0.8 per haul in June–August. Samples were thus too low and variable to consider potential impact of predation on the survival of 0-group cod. As an alternative, the historical beach seine data were analyzed, and when including a much higher number of locations, on average 81 hauls per year, relatively precise estimates of the abundance of larger gadoids could be derived. In that series rather good correspondence is observed between 0- and I-group of the same year-classes (Chapter 2).

Table 3.1 shows the average abundances of potential predators during the period 1919–2001, as well as annual maximum abundances. The only significant negative correlation with 0-group cod abundance was obtained between non-de-trended abundances of 0-group cod and sea trout. However, this correlation broke down when the analysis was conducted on the de-trended data series, suggesting that the trends in 0-group cod and sea trout abundances are negatively correlated rather than there being a direct inter-annual relationship between 0-group cod and sea trout. The positive correlations obtained between

TABLE 3.1 Average Abundance and Annual Maximum in Terms of Catch Per Haul of Various Species Caught in Beach Seine at Locations Sampled More Than 20 Times Between 1919 and 2001 (average 81 Locations. yr^{-1}, the Same Locations shown in Fig. 2.3). The Cross-Correlations (Pearson Coefficients) with Annual Indices of 0-group Cod Abundance for Both Non-De-trended and De-trended Series (third-order Polynomial Regression) are shown in the Last Two Columns. Correlation Analyses were Carried Out on Log-Transformed Data

	Mean	Maximum	Non-de-trended	De-trended
≥I-group cod	1.3	3.7	0.07	−0.03
≥I-group pollack	1.8	9.6	0.01	−0.17
≥I-group saithe	0.4	5.5	0.02	0.00
Sea trout	0.2	0.7	−0.30**	−0.20
0-group cod	13.8	47.4	–	–
0-group pollack	7.6	59.2	0.42***	0.52***
0-group whiting	29.0	85.8	0.14	0.15
0-group saithe	0.8	11.2	0.18	0.18

**$p < 0.01$.
***$p < 0.001$.

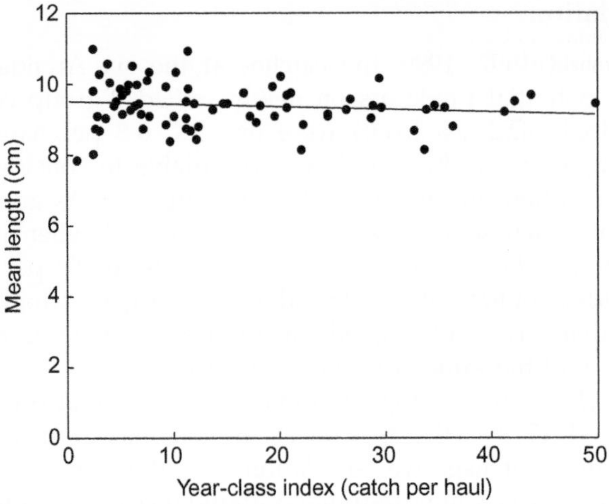

FIGURE 3.4 Relationship between year-class index and average length of 0-group along the Skagerrak coast 1919–2001 (same locations as shown in Fig. 2.4).

the abundances of 0-group cod and 0-group pollack indicate similarity between both trends and inter-annual variability.

Additional information is provided for the Grenlandfjords and Holmestrandfjord (Figs. 2.7 and 2.8) as special cases because severe recruitment collapses happened for the gadoids in those areas in the mid-1960s (Chapter 2). After the severe decline, the abundance of potential predators (i.e., ≥ 20 cm) cod, pollack, saithe, and sea trout collectively decreased by 74% and 65%, respectively, in the two fjords. These abundance changes probably reflect the persistent recruitment failures in the gadoids.

3.3.4 Growth and Year-Class Strength of Cod

Growth curves in settled cod were about the same in all three years (1997–1999) included in the Arendal sites study (Fig. 3.2b). Fig. 3.4 shows the relationship between the annual beach seine abundance of 0-group cod from the entire time series (1919–2001) plotted against the average size of the fish. No relationship between abundance (i.e., year class-index) and size was observed ($p = 0.220$, linear regression). In a more detailed study, Rogers et al. (2011) found minor reductions in growth of 0-group cod at locations with more than 50 cod (e.g., the average length of cod was approximately 0.6 cm smaller at locations with about 250 cod, which are very rare, compared to locations with less than 50 cod). In pollack and whiting too there was no significant

TABLE 3.2 Average Length (cm) of Cod, Pollack, and Whiting Before and After Recruitment Failures in Grenland, Holmestrandfjord, and the Skagerrak coast, and Significance Probabilities Estimated by Two-Sided Student T-test

	Cod			Pollack			Whiting		
	Grenl.	**Holm.**	**Skag.***	**Grenl.**	**Holm.**	**Skag.***	**Grenl.**	**Holm.**	**Skag.***
1953–1966	9.97	9.58	9.56	12.26	12.57	11.88	13.19	12.53	12.57
1966–2001	9.09	9.02	9.25	–	11.66	11.11	12.21	12.02	12.19
Sign. prob.	<0.001	<0.001	<0.001	–	0.004	<0.001	<0.001	<0.001	<0.001

**For the Skagerrak coast periods are 1953–1975 and 1976–2001.*

relationship between year-class strength and size of the fish ($p \geq 0.12$, results not presented).

Table 3.2 provides estimates of the average size of 0-group cod, Pollack, and whiting before and after the recruitment declines in the mid-1960s in the Grenlandfjords and Homestrandfjord. The table also includes corresponding estimates for the gadoids along the entire Skagerrak coast in the mid-1970s, before and after the abrupt reduction in abundances (Chapter 2). In all these subareas, the average sizes of the gadoids were consistently somewhat lower after the recruitment failures. However, the decreases in size fell within the normal inter-annual size variability of all three species (e.g., cod in Fig. 3.4).

3.4 DISCUSSION

3.4.1 Mortality in 0-Group Cod

The initial issue of concern is whether the catch rate patterns observed by beach seine reflect patterns in abundance of newly settled 0-group cod between settling time in June and September when the beach seine estimates appear to correlate positively with abundance at later life stages (Chapter 2). While not exclusively found in littoral and sublittoral habitats sampled by the beach seine, there is observational evidence to suggest that the bulk of the newly settled 0-group in this region is found in such habitats, and that abundance changes after settlement may reflect survival rates and be little influenced by emigration into deeper habitats. There are several studies of cod around the North Atlantic suggesting that the main depth distribution of 0-group cod is <10 m (Methven and Schneider, 1998; Riley and Parnell, 1984), in agreement with the observations from the Skagerrak coast (Chapter 2). Bottom vegetation decrease rapidly with depth, as, for example, eel-grass is rarely found deeper than 6 m in these waters (author's diving observation at 117 beach seine locations). Suitable habitat may therefore be an important reason why the bulk of young-of-the-year cod are found in areas sampled by the beach seine.

The decrease in abundance was most pronounced just after peak settlement, when the fish were only 4–6 cm TL. Substantial active emigration of juveniles of this size range out of habitats offering both ample shelter and suitable food would seem unlikely.

This suggests that it is a reasonable assumption that mortality was the main cause of the decline in abundance of 0-group cod between June and September. If so, the 1997–1999 observations suggest that mortality after settlement is not only high but also variable between years. Processes determining survival patterns from settling until September may contribute very significantly to the variability in subsequent recruitment in Skagerrak coastal waters.

3.4.2 Predation

One of the processes potentially causing variable survival of 0-group cod after settlement might be predation, including cannibalism from older and larger conspecifics. Predation rates were not estimated in this study and would be very hard to obtain, but indirect observation suggests that it is unlikely that predation is the main regulating process underlying the observed abundance changes and mortality patterns. First, the littoral and sublittoral fish community was dominated by other species than the gadoids, in particular gobies. The I-group gobies, of a size similar to newly settled 0-group cod and constituting alternative prey to predators, totally outnumbered cod and other gadoids during the critical period for cod survival. Female gobies spawn by sticking their eggs under small stones and empty shells where they are fertilized and guarded by males until hatching. In the presence of predators the annual gobies take higher risks than gobies spawning over several years (Magnehagen, 1990), and mortality after spawning seems to be a common feature among annual gobies (La Mesa et al., 2005). Therefore, in addition to being highly abundant in June and July, high risk-taking during spawning and reduced condition will probably render annual gobies easy prey to capture for predators.

Several studies have suggested a short-term periodicity in recruitment of 0-group gadoids along the Skagerrak coast, indicating intercohort density dependence with a negative relationship between 0- and I-group cod (Fromentin et al., 1997, 2000). Johannessen (2002) reanalyzed the data and concluded that there is no evidence of short-term periodicity (less than 10 years) in recruitment of 0-group gadoids, nor is there significant density dependent interaction between 0- and I-group cod. The results reported here are in agreement with this conclusion (Table 3.1), that there were no significant correlations between potential predators and the abundance of 0-group cod.

In conclusion, it appears unlikely that predation had significant impact on the different mortality patterns in settled cod (1998 versus 1997 and 1999) or was an important factor for inter-annual variability in the abundance of 0-group cod as measured in September (historical survey).

3.4.3 Growth and Year-Class Strength

3.4.3.1 Food Supply Affecting Survival But Not Growth

The size at settlement and subsequent growth trajectories were quite similar during the three-year study of the Arendal locations from 1997–1999. Hence, different survival rates seemed neither linked to size at settling, nor to growth rate. This is in agreement with the historical data that showed no relationship between year-class strength and size of the 0-group cod (Fig. 3.4). The traditional interpretation of such a lack of relationship is that fish do not experience food limitation, the so-called maximum growth/optimal food condition hypothesis (Karakiri et al., 1991; Veer and Witte, 1993). However, having already rejected predation as the cause of variable mortality in settled 0-group cod, food limitation is nonetheless the most likely explanation. The apparent lack of density-dependent growth and food limitation as the cause of variable mortality might initially appear as conflicting conclusions. However, there is a solution to this problem.

The lack of a relationship between the year-class strength and growth can also arise from a situation where food availability always limits growth. Under this interpretation, intra-specific competition is independent of year-class strength. In years with ample food supply, competition for food is kept at the same level as in years with poor food supply by a larger number of cod surviving: hence, food supply affects the number of survivors and not growth. Theoretical analyses as a background for this scenario are presented in the following section (which may be skipped).

3.4.3.2 Theoretical Analyses of Settlement, Growth, and Survival

Theoretically, there are two possible scenarios that might lead to year-class strength being independent of the average size of the fish, one without intraspecific competition for food and one with competition. The conceptual models in Fig. 3.5a and b explore the relationship between survival and growth with changing prey densities under the two scenarios. Below a specific prey density (p_{min}) no cod will survive

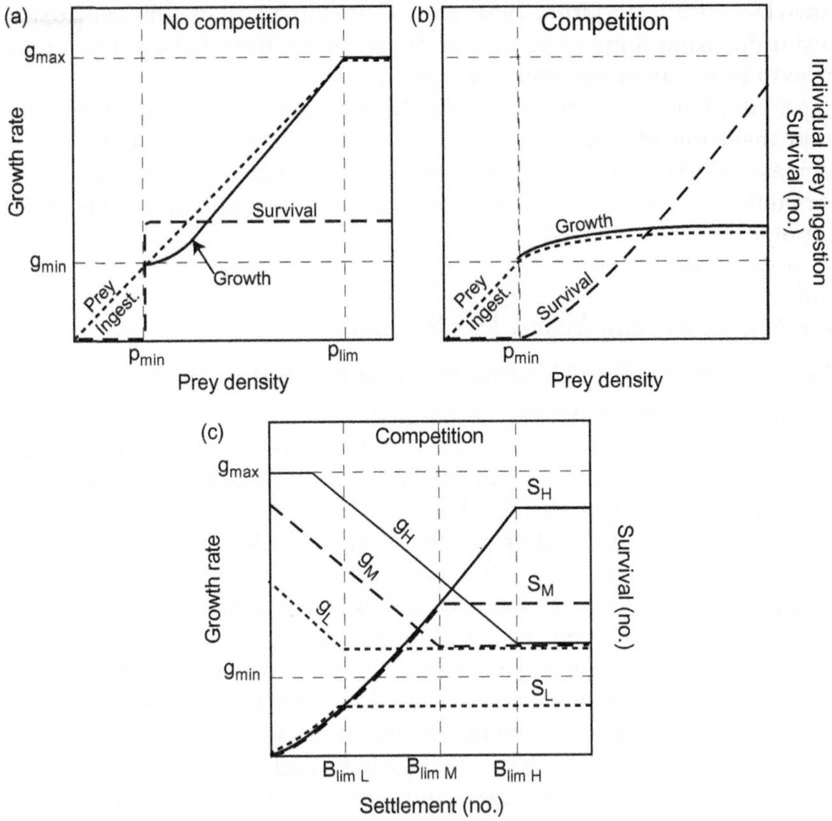

FIGURE 3.5 Theoretical relationship between prey density and growth rate and survival, (a) without competition for food; (b) with competition for food; (c) theoretical relationship between settlement in numbers below and above the capacity of the benthic habitat and growth and survival, in situation with low (B_{limL}), medium (B_{limM}), and high (B_{limH}) prey densities; p_{min}—minimum prey concentration for survival, g_{min}—growth rate associated p_{min}, g_{max}—maximum growth rate in 0-group cod, p_{lim}—prey density associated with g_{max}, g_L, and S_L growth and survival associated with B_{limL}. See text for explanation.

(this is obviously a simplification because there will probably always be some cod that will find sufficient food because of patchy distribution of prey). This minimum prey density is associated with a minimum growth rate (g_{min}). Accordingly, above a specific prey concentration (p_{lim}) growth rate will be maximal (g_{max}). Above minimum prey density (p_{min}) there will be enough food for all individuals as long as there is no competition, and the survival will therefore not be affected by increasing prey density (Fig. 3.5a). Theoretically, no competition under low prey density is possible if the abundance of cod never becomes sufficiently high to significantly impact prey abundance.

Growth rate, on the other hand, will increase with increasing prey density until maximum growth rate is reached. The relationship between growth rate and prey density will not be linear because extra energy is required to obtain food at low prey densities. Without competition, food ingestion of the individual cod will increase proportionally with increasing prey concentrations until p_{lim} is reached. Prey densities above p_{lim} correspond to the maximum growth/optimal food condition hypothesis.

In the case of intraspecific competition for food (Fig. 3.5b) survival will increase with increasing prey density, but because of a higher number of survivors the individual cod will not ingest more food with increasing prey densities, nor will growth increase notably (hence, in this case there is no p_{lim}). Theoretically, under this scenario, growth rate should be minimal (g_{min}). However, because of inhomogeneous distribution of prey some individuals will ingest more prey than minimum for survival just by chance, whereas some will ingest more than others because they are better competitors. Consequently, growth rate will end up somewhere between g_{min} and g_{max}, probably closer to g_{min}.

Both of these scenarios will result in there being no relationship between the abundance (survival) and the average size of the fish. In the case of no competition, all cod will have sufficient food to survive at prey densities above p_{min}. In this situation, mortality will not be regulated by food limitation, and the number of settlers would be expected to be in close agreement with recruitment as measured in September. This is, however, in conflict with the results obtained during the 1997–1999 study in which both high and variable mortality rates were observed, resulting in variable survival in September despite similar abundance at settling. On the other hand, the scenario with higher survival rates with increasing food availability provides a reasonable explanation for the variable survival rates and the lack of relationship between year-class strength and size in 0-group cod (Fig. 3.4). In agreement with this, Lekve et al. (2002), who studied skewness in the length distributions of cod from the Skagerrak coast, found evidence of competition in 0-group cod.

In the three years 1997–1999, relatively high mortality in cod was observed shortly after peak settlement, suggesting that the numbers settling were higher than what the available food resources could support. However, a three-year study is obviously not sufficient to assess whether this is the rule. A scenario in which the number settling is lower than the carrying capacity of the habitat should therefore be explored. This might happen if the pelagic juvenile cod when settling shift from feeding on plankton in low concentrations to more abundant benthic and hyperbenthic prey after settling.

In Fig. 3.5c the impact on abundance and growth is analyzed in relation to increasing settlement in situations in which the benthic capacity (B_{lim}) is low, medium, and high, which correspond to low, medium, and high prey densities. As the scenario with no competition appears unlikely, only patterns under a competition scenario are explored. Below B_{lim} there will be no mortality due to food limitation and the abundance will increase approximately proportional to settlement. Growth rate, on the other hand, will decrease due to increased competition as settlement approaches B_{lim}. Settlement above B_{lim} will not affect growth rate or survival. However, as the prey density increases from low concentrations (B_{limL}) to high concentrations (B_{limH}), survival will increase but not growth rates. Settlement above B_{lim} corresponds to the scenario in Fig. 3.5b. Obviously, the chance of settling in numbers below B_{lim} will decrease with increasing settlement. Therefore, if settlement below the capacity of the habitat were common, one would expect the size of cod in poor year-classes to be larger than those in strong year-classes. However, there is no evidence of such a relationship (Fig. 3.4). Therefore, settlement below the carrying capacity of the benthic habitat appears to be rare.

3.4.3.3 *Growth of 0-Group Cod in Excess of Food*

The preceding proposed mechanisms underlying the lack of relationship between year-class strength and size in 0-group cod predicts that growth rate of cod in its natural environment is well below maximum growth rate of the species. In agreement with this, laboratory experiments have shown that reared cod can grow much faster than normally observed in wild populations (Braaten, 1984; Otterlei et al., 1999). Furthermore, there is some historical evidence to suggest that under exceptional events resulting in unusually high prey densities, wild cod too can grow substantially faster than what is normally observed. The historical beach seine survey series show that in 1938 there was a unique event in the Inner Oslofjord with extraordinary high abundances of 0-group cod (71.6 per haul as opposed to the 1936–2001 average of 3.7 per haul; Fig. 2.9a). The juveniles occurring in that single rich year were not only numerous but also exceptionally big, 12.3 cm–16.1 g contrasting with the long-term average of 9.4 cm–7.0 g (estimated weight of average length fish). Dannevig (1949a, 1949b) monitored the growth of the 1938 year-class by additional sampling. From July to October the average catch in the beach seine decreased only from 62 to 53 cod per haul (based on 27 beach seine locations), suggesting low mortality. Dannevig (1949b) studied individual growth rates in these cod on the basis of sclerite increments in scales and concluded that individual growth rates were in agreement with that of the population (population growth rate may be affected by size-selective mortality).

The low mortality and high growth rates in 0-group cod this year concurred with advection of high concentrations of the copepod *Calanus finmarchicus* into the fjord during early summer (Wiborg, 1940).

After the collapse of the gadoids in the Grenlandfjords and Holmestrandfjord in the mid-1960s, the size of the gadoids decreased slightly (Table 3.2). A newly settled cod of 4 cm weighs 0.5 g, whereas a 10 cm cod weighs 8.5 g; hence the gain in weight from settling to September is very significant. Had feeding conditions prior to the collapse prevailed, the opposite of a decrease in size would have been expected. As shown in Chapter 2, the abundance of young-of-the-year gadoids was reduced by ≥ 90%. Furthermore, the abundance of non-gadoid fishes also decreased in those areas (Fig. 2.10). The much reduced numbers of juvenile fish would be expected to grow faster and attain greater sizes in September. The opposite happened, and a possible and likely explanation is that food supply was not maintained at previous levels but decreased significantly and sufficiently to reduce survivorship in many juvenile fish species. Interpreted in light of the conceptual model in Fig. 3.5c, the food supply (B_{lim}) declined substantially and reduced survival rates to a major degree (but had only slightly negative impact on growth rates).

3.4.4 A Recruitment Hypothesis for Cod

In the Inner Oslofjord, the extraordinary high abundance and size of the 0-group cod in 1938 concurred with advection of high concentrations of the calanoid copepod *C. finmarcichus* into the fjord (Wiborg, 1940). In the North Sea, Beaugrand et al. (2003) found that a marked reduction in recruitment of cod in the 1980s coincided with a decrease in the size of calanoid copepods, a phenomenon that was particularly pronounced in July. Rothschild (1998) concluded that high abundance of either *Calanus* or *Paracalanus/Pseudocalanus* appeared to be required to produce a large year class of cod in the North Sea. Hence, also from other areas there is evidence to suggest that pelagic prey is important for the recruitment of cod, and that the copepods *C. finmarchicus* and *Paracalanus/Pseudocalanus* are particularly important prey (Heath and Lough, 2007; Robert et al., 2011; Sundby, 2000).

Referring to information provided in Chapter 2, the incidents of significant and persistent recruitment failure in the semi-enclosed Inner Oslofjord occurred during a period with increasing nutrient loads and at a time when the concentrations of other pollutants were still low. Strong similarities were observed between the incident in Inner Oslofjord and other subareas of the Skagerrak coast, and eutrophication was the only factor common to the various subareas. Increasing eutrophication occurred immediately prior to the declines in recruitment, and it was concluded that the recruitment collapses were most likely somehow caused by the gradually increasing nutrient loads. Elevated nutrient concentrations as well as altered nutrient composition will affect competition in phytoplankton (Anderson et al., 2002; Smayda 1990), and given the gadoid's documented strong dependence on planktonic prey such as calanoid copepods, it may be hypothesized that one effect of enhanced nutrient loads was abrupt shifts in the plankton community that deprived young-of-the-year gadoids of their natural prey (diet in cod during the critical recruitment phase is investigated in Chapter 4).

When settling, cod shift from a spacious pelagic habitat to a much smaller depth-limited benthic habitat along the shore. If cod continue to feed on planktonic prey after settlement, it seems reasonable to assume that competition for food will increase substantially after settlement and thus provide for a mechanistic explanation as to why cod seem to settle in surplus of the carrying capacity of the benthic habitat (B_{lim}, Fig. 3.5c) and why mortality rates appear to be particularly high shortly after peak settlement.

Based on the information and logic presented, the following recruitment hypothesis for cod is proposed:

> The survival of 0-group cod is limited by food availability after settlement (3 to 5 months old), and recruitment variability results from differences in food supply due to inter-annual variability in the energy flow pattern at low trophic levels of the pelagic food web.

This hypothesis recognises the early "critical period" thinking but suggests that rather than maintaining the strong emphasis on the pelagic larval period as most significant, greater attention should be directed at food-web processes during the first months after settling into the benthic environment, but when cod is still predominantly a planktivore. This "energy flow" hypothesis may explain both natural recruitment variability and what appears as more exceptional events such as the eutrophication-induced recruitment collapses of cod and the other gadoids (Chapter 2). It postulates that food availability shows high inter-annual variability and that gradually increasing nutrient loads may result in abrupt and persistent food deprivation of the 0-group gadoids. Comprehensive tests of the recruitment hypothesis in the field are presented in Chapter 4.

References

Anderson, J.T., 1988. A review of size dependent survival during pre-recruit stages of fishes in relation to recruitment. J. Northw. Atl. Fish. Sci. 8, 55–66.

Anderson, D.M., Glibert, P.M., Burkholder, J.M., 2002. Harmful algal blooms and eutrophication: nutrient sources, composition, and consequences. Estuaries 25, 704–726.

Beaugrand, G., Brander, K.M., Lindley, J.A., Souissi, S., Reid, P.C., 2003. Plankton effect on cod recruitment in the North Sea. Nature 426, 661–664.

Blegvad, H., 1917. Om fiskenes føde i de danske farvande innden for Skagen. Beretning til Landbrugsministeriet 24, 17–72 (in Danish).

Blaxter, J.H.S. (Ed.), 1974. The Early Life History of Fish. Springer, Berlin.

Blaxter, J.H.S., Gamble, J., Westernhagen, H. (Eds.), 1989. The early life history of fish. Rapp. Cons. int. Explor. Mer 191

Bogstad, B., Lilly, G.R., Mehl, S., Pálsson, O.K., Stefánsson, G., 1994. Cannibalism and year-class strength in Atlantic cod (*Gadus morhua* L.) in Arcto-boreal ecosystems (Barents Sea, Iceland, and eastern Newfoundland). ICES Mar. Sci. Symp. 198, 576–599.

Braaten, B., 1984. Growth of cod in relation to fish size and ration level. Flødevigen Rapportser. 1, 677–710.

Dannevig, A., 1949a. The variation in growth of young codfishes from the Norwegian Skagerrak coast. Fiskeridir. Skr. Ser. Havunders. 9, 1–12.

Dannevig, A., 1949b. Cod scales as indicator of local stocks. Fiskeridir. Skr. Ser. Havunders. 9, 13–22.

Fabre-Domergue, P., Bietrix, E., 1897. La période critique post-larvaire des poisons marins. Bull. Mus. Nat. Hist. Nat. Paris 3, 57–58.

Fromentin, J.M., Stenseth, N.C., Gjøsæter, J., Bjørnstad, O.N., Falck, W., Johannessen, T., 1997. Spatial patterns of the temporal dynamics of three gadoid species along the Norwegian Skagerrak coast. Mar. Ecol. Prog. Ser. 155, 209–222.

Fromentin, J.M., Gjøsæter, J., Bjørnstad, O.N., Stenseth, N.C., 2000. Biological processes and environmental factors regulating the dynamics of the Norwegian Skagerrak cod populations since 1919. ICES J. Mar. Sci. 57, 330–338.

Gotceitas, V., Fraser, S., Brown, J.A., 1997. Use of eelgrass beds (*Zostera marina*) by Juvenile Atlantic cod (*Gadus morhua*). Can. J. Fish. Aquat. Sci. 54, 1306–1319.

Grant, S.M., Brown, J.A., 1998a. Nearshore settlement and localized populations of age 0 Atlantic cod (*Gadus morhua*) in shallow coastal waters of Newfoundland. Can. J. Fish. Aquat. Sci. 55, 1317–1327.

Grant, S.M., Brown, J.A., 1998b. Diel foraging cycles and interactions among juvenile Atlantic cod (*Gadus morhua*) at a nearshore site in Newfoundland. Can. J. Fish. Aquat. Sci. 55, 1307–1316.

Healey, M.C., 1971. Gonad development and fecundity of the sand goby, *Gobius minutus* Pallas. Trans. Am. Fish. Soc. 3, 520–526.

Heath, M.R., Lough, R.G., 2007. A synthesis of large-scale patterns in the planktonic prey of larval and juvenile cod (*Gadus morhua*). Fish. Oceanogr. 16, 169–185.

Hesthagen, I.H., 1977. Migrations, breeding and growth in *Pomatoschistus minutus*. Sarsia 63, 17–26.

Hjort, J., 1914. Fluctuations in the great fisheries of northern Europe viewed in the light of biological research. Rapp. P.-v. Réun. Cons. int. Explor. Mer 20, 1–228.

Johannessen, T., 2002. Is there short-term periodicity in gadoid recruitment along the Norwegian Skagerrak coast? Mar. Ecol. Prog. Ser. 241, 227–229.

Karakiri, M., Berghahn, R., van der Veer, H.W., 1991. Variations in settlement and growth of 0-group plaice (*Pleuronectes platessa* L.) in the Dutch Wadden Sea as determined by otolith microstructure analysis. Neth. J. Sea Res. 27, 345–351.

Kendall, A.W., Duker, G.J., 1998. The development of recruitment fisheries oceanography in the United States. Fish. Oceanogr. 7, 69–88.

Lasker, R., Sherman, K., 1981. The early life history of fish: recent studies. Rapp. Cons int. Explor. Mer 178.

Lekve, K., Ottersen, G., Stenseth, N.C., Gjøsæter, J., 2002. Length dynamics in juvenile coastal Skagerrak cod: effects of biotic and abiotic processes. Ecology 83, 1676–1688.

Magnehagen, C., 1990. Reproduction under predation risk in the sand goby, *Pomatoschistus minutus*, and the black goby, *Gobius niger*: the effect of age and longevity. Behav. Ecol. Sociobiol. 26, 331–335.

May, R.C., 1974. Larval mortality in marine fishes and the critical period concept. In: Blaxter, J.H.S. (Ed.), The Early Life History of Fish. Springer, Berlin, pp. 3–19.

Mesa, M.L., Arneri, E., Caputo, V., Iglesias, M., 2005. The transparent goby, *Aphia minuta*: review of biology and fisheries of a paedomorphic European fish. Rev. Fish. Biol. Fisher. 15, 89–109.

Methven, D.A., Schneider, D.C., 1998. Gear-independent patterns of variation in catch of juvenile Atlantic cod (*Gadus morhua*) in costal habitats. Can. J. Fish. Aquat. Sci. 55, 1430–1442.

Mori, S., Nagoshi, M., 1987. Growth and maturity size of the three-spined stickleback *Gasterosteus aculeatus* in rearing pool. Bull. Fac. Fish., Mie Univ. 14, 1–10.

Otterlei, E., Nyhammer, G., Folkvord, A., Stefansson, S.O., 1999. Temperature and size dependent growth of larval and juvenile cod (Gadus morhua L.) - a comparative study between Norwegian coastal cod and Northeast Arctic cod. Can. J. Fish. Aquat. Sci. 56, 2099–2111.

Riley, J.D., Parnell, W.G., 1984. The distribution of young cod. Flødevigen Rapportser. 1 (1984), 563–580.

Robert, D., Levesque, K., Gagné, J.A., Fortier, L., 2011. Change in prey selectivity during the larval life of Atlantic cod in the southern Gulf of St Lawrence. J. Plankton Res. 33, 195–200.

Rogers, L.A., Stige, L.C., Olsen, E.M., Knutsen, H., Chan, K.-S., Stenseth, N.C., 2011. Climate and population density drive changes in cod body size throughout a century on the Norwegian coast. Proc. Natl. Acad. Sci. USA 108, 1961–1966.

Rothschild, B.J., 1998. Year class strength of zooplankton in the North Sea and their relation to cod and herring abundance. J. Plankton Res. 20, 1721–1741.

Shepherd, J.G., Cushing, D.H., 1980. A mechanism for density-dependent survival of larval fish as the basis of a stock-recruitment relationship. J. Const. Int. Explor. Mer 39, 160–167.

Smayda, T., 1990. Novel and nuisance phytoplankton blooms in the sea: evidence for a global epidemic. In: Graneli, E., Sundstrom, B., Edler, L., Anderson, D.M. (Eds.), Toxic Marine Phytoplankton. Elsevier, New York, pp. 29–40.

Sokolowska, E., Kulczykowska, E., 2006. Annual reproductive cycle in two free living populations of three-spined stickleback (*Gasterosteus aculeatus* L.): patterns of ovarian and testicular development. Oceanologia 48, 103–124.

Sundby, S., 2000. Recruitment of Atlantic cod stocks in relation to temperature and advection of copepod populations. Sarsia 85, 277–298.

Tupper, M., Boutilier, R.G., 1995. Size and priority at settlement determine growth and competitive success of newly settled Atlantic cod. Mar. Ecol. Prog. Ser. 118, 295–300.

Van der Veer, H.W., Witte, J.I.J., 1993. The "maximum growth/optimal food condition" hypothesis: a test for 0-group plaice *Pleuronectes platessa* in the Dutch Wadden Sea. Mar. Ecol. Prog. Ser. 101, 81–90.

Ware, D.M., 1975. Relation between egg size, growth, and natural mortality of larval fish. Can. J. Fish. Aquat. Sci. 32, 2503–2512.

Wheeler, A., 1969. The Fishes of the British Isles and North-west Europe. MacMillian, London.

Wiborg, K.F., 1940. The production of zooplankton in the Oslofjord 1933–1934. Hvalråd. Skr. Sci. Results Mar. Biol. Res. 21, 1–87.

Zar, J.H., 1974. Biostatistical Analysis. Prentice-Hall, Englewood Cliffs.

CHAPTER

4

Growth and Mortality in Settled Atlantic Cod in Relation to Diet—Evidence for a Recruitment Mechanism

Tore Johannessen[1], *Ingrid Berthinussen*[2], *Jens-Otto Krakstad*[3] *and Anders Fernö*[4]

[1]Institute of Marine Research, Flødevigen, Norway [2]Norwegian Polar Institute, Tromsø, Norway [3]Institute of Marine Research, Bergen, Norway [4]University of Bergen, Department of Biology, Bergen, Norway

4.1 INTRODUCTION

4.1.1 Empirical Background

A long-term times series of annual beach seine sampling (since 1919) along the Norwegian Skagerrak coast has revealed repeated incidents of abrupt and persistent local recruitment collapses in gadoid fishes in relation to increasing eutrophication (Chapter 2). In the various affected areas, the collapses occurred simultaneously in the gadoids, and after the collapses there have been no signs of recovery (e.g., since 1930 in Inner Oslofjord). The abundance of other littoral fishes is also generally much lower in these areas. A number of hypotheses regarding the underlying mechanisms were considered unlikely, such as, overfishing, low oxygen, habitat degradation through changes in bottom flora, and direct impact of contaminants on fish and spawning products (Chapter 2). One problem in disentangling the mechanism behind these recruitment collapses is a general lack of knowledge about which

From an Antagonistic to a Synergistic Predator Prey Perspective.
DOI: http://dx.doi.org/10.1016/B978-0-12-417016-2.00004-X

factors limit recruitment and generate inter-annual recruitment variability in the gadoids, and in fish in general (see Chapter 3 and references therein).

4.1.2 Fish Recruitment

Because historical data are often readily available, most recruitment studies in fishes have essentially constituted analyses of the relationships between physical (wind, currents, temperature, etc.) and/or biological (growth rates, predators, prey, etc.) variables and recruitment variability for a given species (e.g., Beaugrand et al., 2003; Johannessen and Tveite, 1989; Rothschild, 1998). This was the approach during the first part of this study as well (Chapter 2). Smith (1994; cited from Kendall and Duker, 1998) states, "Since the 1920s, correlations of the strength of year-classes with environmental factors....began to take a certain melancholy consistency. Initial data might suggest a high correlation...but eventually the correlation would fail." Indeed, because of high inter-correlations between physical and biological variables, correlation studies can only provide hypothesis of causal relationships. This is also the case with laboratory studies of recruitment mechanisms because they can never claim to fully mimic what happens in nature. So, 100 years after Hjort (1914) proposed the first recruitment hypothesis, we still have not resolved the recruitment puzzle. The lack of success is strong evidence suggesting that neither correlation studies nor laboratory experiments have been adequate methods. Therefore, in order to confirm recruitment hypotheses generated by such studies, the proposed mechanisms have to be tested in the field.

There is consensus that most of the recruitment variability occurs during early life stages of the fish. Although more recently there has been growing interest in older juvenile stages (Houde, 2008), most of the research into the recruitment puzzle has been directed toward larval and post-larval stages (Anderson, 1988), which continue to receive much attention (e.g., Burrow et al., 2011; Kristiansen et al., 2011). Unfortunately, here we are apparently faced with a Gordian knot. Testing recruitment hypotheses during larval and post-larval stages seems to be almost unresolvable, probably because mortality rates are very high, which render precise sampling very difficult. The sampling problem is particularly evident in oceanic fish stocks in which the larvae are spread over wide areas.

4.1.3 Field Test of a Recruitment Hypothesis for Cod

Due to the inherent problem of studying mortality rates during the pelagic phase, the focus of a three-year recruitment study at locations near Arendal (referred to in Chapter 3) was the period from settlement to the year-class strengths mainly determined in these waters, that is, by September at the 0-group stage (Chapter 2). Based on this three-year study, Atlantic cod (*Gadus morhua*) was chosen as the model species to further test the mechanisms behind recruitment variability and the observed recruitment collapses.

During all three years peak abundance in newly settled cod occurred around mid-June, and the numbers were about the same each year. Growth rates from settlement until September were quite similar too. However, different mortality rates shortly after peak abundance resulted in one strong year-class and two weak year-classes, suggesting that processes after settlement rather than during the larval phase were most significant in the regulation of year-class abundance. Based on analyses of the theoretical relationship between number of cod settling, growth rates, mortality, and year-class strength, it was suggested that food availability after settlement limits survival of cod in these waters.

Theoretical considerations in combination with empirical data suggested that cod generally settle in surplus of the carrying capacity of the benthic environment they colonize. It is likely, however, that settled juveniles maintain a planktivorous feeding mode, and several correlation studies have indicated that copepods are important for the recruitment in Atlantic cod (Heath and Lough, 2007; Rothschild, 1998; Sundby, 2000). Based on these observations, the following hypothesis was proposed: *The survival of 0-group cod is limited by food availability after settlement, and recruitment variability results from differences in food supply arising from inter-annual variability in the energy flow pattern at low trophic levels of the pelagic food web.*

In this chapter, the following two predictions of the hypothesis were tested in a comparative field study:

1. The number of cod that settle does not limit recruitment; hence, recruitment is generally decoupled from the larval and post-larval phase.
2. Inter- and intra-annual variability in the number of cod surviving through the critical period are related to food conditions in terms of quantity and quality of the prey (the food conditions vary between years and between subareas with and without severe recruitment collapses).

Comprehensive studies of post-settled cod along the coasts of the northwest Atlantic have been carried out, investigating aspects such as settlement (Grant and Brown, 1998a; Methven and Bajdik, 1994; Tupper and Boutilier, 1995a, 1995b), site fidelity (Grant and Brown, 1998a), habitat preferences (Gotceitas et al., 1997; Grant and Brown, 1998a; Tupper and Boutilier, 1995b), and nourishment (Grant and Brown, 1998b, 1999).

However, none of the studies from the northwest Atlantic have included inter-annual variability in mortality rates in post-settled cod and how this may affect recruitment.

4.2 METHODS

4.2.1 Sampling

This study was conducted in the Grenlandfjords (Fig. 4.1), where recruitment of gadoids declined abruptly to unprecedented low levels from the mid-1960s onwards, and in the Risør area, where less severe recruitment declines have been observed (Chapter 2). In Grenland, 10 locations were sampled. Around Risør, three subareas with several seining locations each were included in the studies (Fig. 4.1): Sandnesfjord (eight locations), Søndeledfjord (seven locations), and Risør archipelago (four locations). Søndeledfjord is a semi-enclosed fjord system, Sandnesfjord more open, and the Risør archipelago relatively exposed toward Skagerrak. We decided to distinguish between the different Risør subareas because short-term variability in 0-group cod recruitment along the Skagerrak coast is spatially structured at the scales of fjords (Fromentin et al., 1998).

Sampling was carried out with a beach seine identical to that used in the annual coastwide survey carried out since 1919 (Chapter 2), except that during sampling periods prior to the regular survey in September we used a seine with finer mesh—10 mm rather than 15 mm stretched meshes. A fine mesh was selected to meet the objective of retaining also the newly settled gadoid juveniles in the months prior to the annual survey, always carried out in September. By September, earlier studies have determined that 0-group gadoids are big enough to be fully retained in the 15 mm mesh seine (Tveite, 1971). In 1995, sampling was carried out during the last week of June and the last week of September. In 1996–1998, sampling occurred in the first week of July, in mid-August, and during the last week of September.

The main target species of this study was Atlantic cod, but the fine mesh seine also samples a range of other co-occurring fish and invertebrate species for which abundance data were also estimated from the July and August surveys.

4.2.2 Diet and Condition

The ungutted specimens of cod were put on ice immediately after sampling, then measured (total length of up to 50 per location, precision 1 mm), weighed (W, 0.01 g), and frozen pending further investigations.

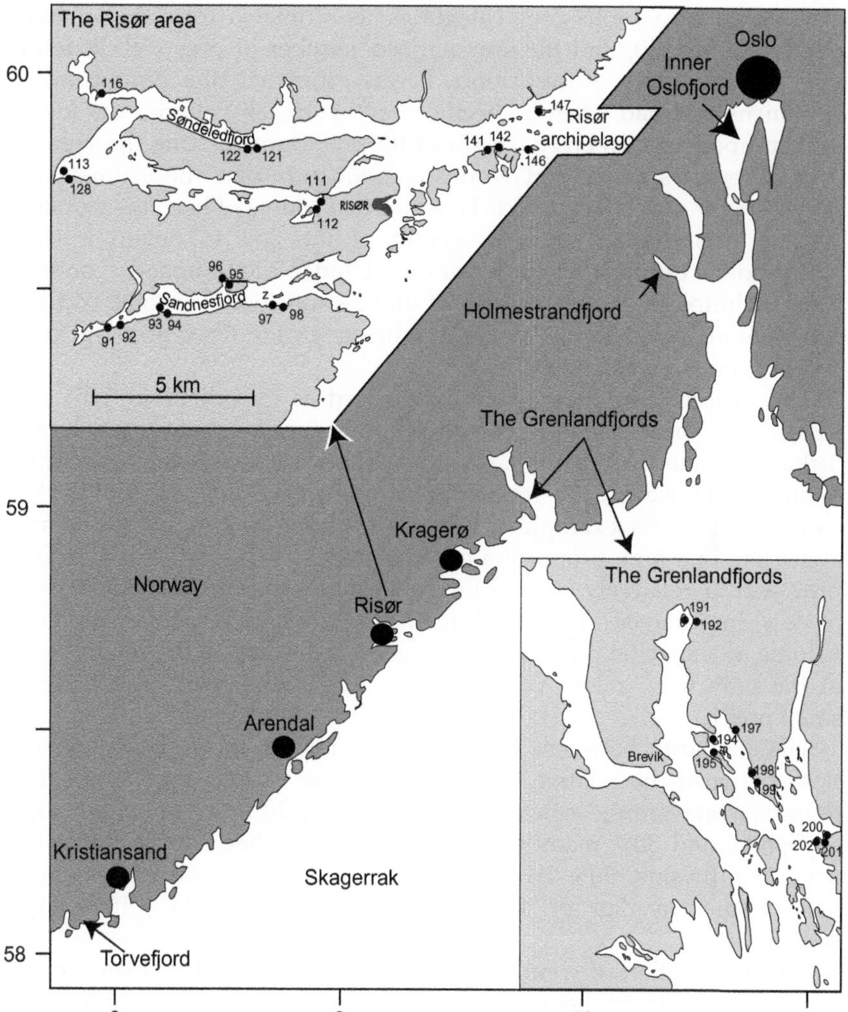

FIGURE 4.1 The Norwegian Skagerrak coast. Beach seine sampling was carried out in the Risør area and in the Grenlandfjords.

After thawing, the stomach contents (S) and liver (L) were weighed (0.01 g). Relative weight of stomach content was estimated as Si = (S/W)*100%, and relative liver weight as Li = (L/W)*100%. Prey organisms were identified to the lowest possible taxa and counted. Relative volume by taxon was estimated according to Gibson and Ezzi (1987). Comparison of the diet between years and locations was based on the relative contribution to the time- and location-specific pooled stomach contents of the various prey categories, as estimated from the summarized volumes (V), and the summarized number of prey (N). In addition, the frequency of occurrence F was estimated (the proportion of specimens that had eaten a specific prey). Stomach volume was found to be proportional to body weight of the cod. Summarizing the volume over different sizes of cod would consequently bias the estimates of prey in favor of the larger cod. To avoid this source of bias, the stomach volume was divided by the weight of the cod to obtain a size-independent estimate of prey volume. Unidentified stomach contents were excluded from the analyses of diet composition. Up to 30 cod per site were analyzed from Grenland, whereas in the Risør area up to 10 cod per site were analyzed.

The various prey types were categorized as hyperbenthic (HB) and pelagic (P). There was also a small proportion of benthic prey that was included in the hyperbenthic category. The main focus was on volume (biomass). However, the frequency of occurrence may be useful as a correction tool, for example; a prey contributing to a high volume and appearing in a high proportion of the stomachs is important, whereas a prey with a relatively high volume but appearing in few stomachs is of less importance. The combination of numerical proportion and volume of a specific prey provides a good indication of the relative size of the prey (e.g., a low volume and a high number of prey indicate small prey).

The similarity between diets was estimated as the sum of intersections between the relative volume (V) of the various prey types. For example, one sample consisting of 40% fish, 20% mysids, 10% large copepods, and 30% medium copepods and another sample with 10% fish, 25% mysids, 50% large copepods, and 15% cladocerans would have a diet overlap of 40% (10% fish + 20% mysids + 10% large copepods).

Characterization of condition was based on the relative liver weight estimates. However, because of accidental thawing of a sample from Grenland in July 1996, livers could not be extracted from a subset of specimens. Comparisons of the condition of cod from Grenland and Risør in July 1996 were thus made by the less sensitive Fulton condition factor (W/L^3, estimated from the fresh samples that were more numerous than the samples collected for diet analyses). Because the post-settled cod did not conform to the assumption of isometric growth (i.e., condition factor is independent of length), the comparisons were made within 5 mm length intervals (except 60–69 mm, Fig. 4.6d). In addition, as an indication of the condition of cod in Grenland in 1996, the liver index was estimated indirectly using the linear relationship between the Fulton index and the liver index in individual cod from Risør in July 1996.

4.2.3 Statistical Analyses

The relationship between abundances (catch per beach seine haul), mean lengths, and year-class index as measured in September (Chapter 2) was studied by Pearson correlation analyses. To avoid potential influences of local differences (e.g., by the average size of cod varying between subareas), mean values of catches and lengths were in some occasions standardized to zero mean per subarea.

Inter- and intra-annual variability in the average size of the cod was estimated in terms of mean deviation (MD), which is more intuitive and less influenced by outliers than standard deviation (SD); MD is estimated on the basis of $|x_i - \overline{x}|$ while in SD this difference is squared, $(x_i - \overline{x})^2$. Inter-annual and seasonal MD were estimated as follow:

$$MD = \frac{1}{16}\sum\nolimits_{i=1}^{4}\sum\nolimits_{j=1}^{4}|l_{ij} - \bar{l}_i| \tag{4.1}$$

In the case of inter-annual MD, i represents the four subareas (Sandnesfjord, Søndeledfjord, Risør archipelago, and Grenland), j the four years (1995–1998), l_{ij} the average size of cod in subarea i in year j, and $\bar{l}_i$ the average size of cod in subarea i during 1995–1998. In the case of intra-annual MD, i represents the subareas while j represents the years. Differences between MDs were tested using Levine's test. Because both inter- and intra-annual MDs involved estimation of four means per sample, the degrees of freedom were reduced by 8 (4×2).

Chi-square contingency analysis was used to test whether the frequency of occurrence of prey types was different between years and subareas. Some prey categories may be unimportant in terms of biomass but may appear in a high number of stomachs. To reduce the impact of such "unimportant" prey, categories contributing to less than 2% of stomach volume were excluded from the analysis of frequency of occurrence (but prey $\geq$2% in one group and 0% in another are included).

In some cases in which the samples did conform to the underlying assumption of normality, nonparametric methods were applied.

4.3 RESULTS

Based on evaluations of possible causes of reduced abundances in 0-group cod after settlement observed during the Arendal study in Chapter 3, it was concluded that the main cause was mortality rather than migration. Except for less frequent sampling (approximately 1.5 months versus biweekly), there were no methodological differences between this study and the Arendal study. Therefore, reduction in abundances after settlement observed in the present study is also assumed to reflect mortality.

4.3.1 Abundance, Size at Settlement, and Growth in 0-Group Cod

The abundance of 0-group cod tends to vary considerably between sampling locations, probably due to habitat differences. Because one subarea may contain locations with more suitable habitat than another subarea, attention will be on relative changes in abundance rather than the absolute differences between subareas. To put the results into perspective, it should be noted that the average September survey catch of 0-group cod across the period 1919–2001 and all locations along the Skagerrak coast is 15.3 specimens per haul.

In the subareas around Risør the abundance of 0-group cod decreased substantially from July onwards (Fig. 4.2a–c), with the largest decline taking place between July and mid-August. Between August and September there were only minor reductions, with good correspondence between the estimates ($r = 0.98$, $p < 0.001$, $n = 9$), that is, high abundance in August resulted in high abundance in September. In contrast, there was relatively poor correspondence between abundances in July and September ($r = 0.4$, $p = 0.21$, $n = 12$). This was particularly evident in Sandnesfjord, where rich settlement in 1995 was followed by a 95% reduction, resulting in an average year-class in September, whereas in 1998 substantially lower settlement resulted in a strong year-class. In August 1997, the abundance was probably slightly underestimated due to high surface temperatures (average 20.1°C in July–August, daily measurements at 1 m depth) that may have caused cod to temporarily seek deeper and colder water. In September the temperature was back to normal (14.9°C, mean 15–30 September).

Good correspondence was observed between the year-class strength of cod as measured in September in Sandnesfjord and Risør archipelago and those along the Skagerrak coast (Skagerrak coast data obtained from Fig. 2.4a; $r = 0.95$ and $r = 0.78$, respectively; Pearson correlation on log-transformed annual year-class indices): a weak year-class appeared in 1997, relatively strong year-classes in 1995 and 1998, and a strong year-class in 1996 (Fig. 2.4a). In contrast, in Søndeledfjord a relatively strong year-class appeared in 1995, an average year-class in 1997, whereas the 1996 and 1998 year-classes were both relatively weak ($r = 0.08$ between Søndeledfjord and the Skagerrak coast). In agreement with this, there has been low correspondence between year-class strength in Søndeledfjord and Risør archipelago in recent years

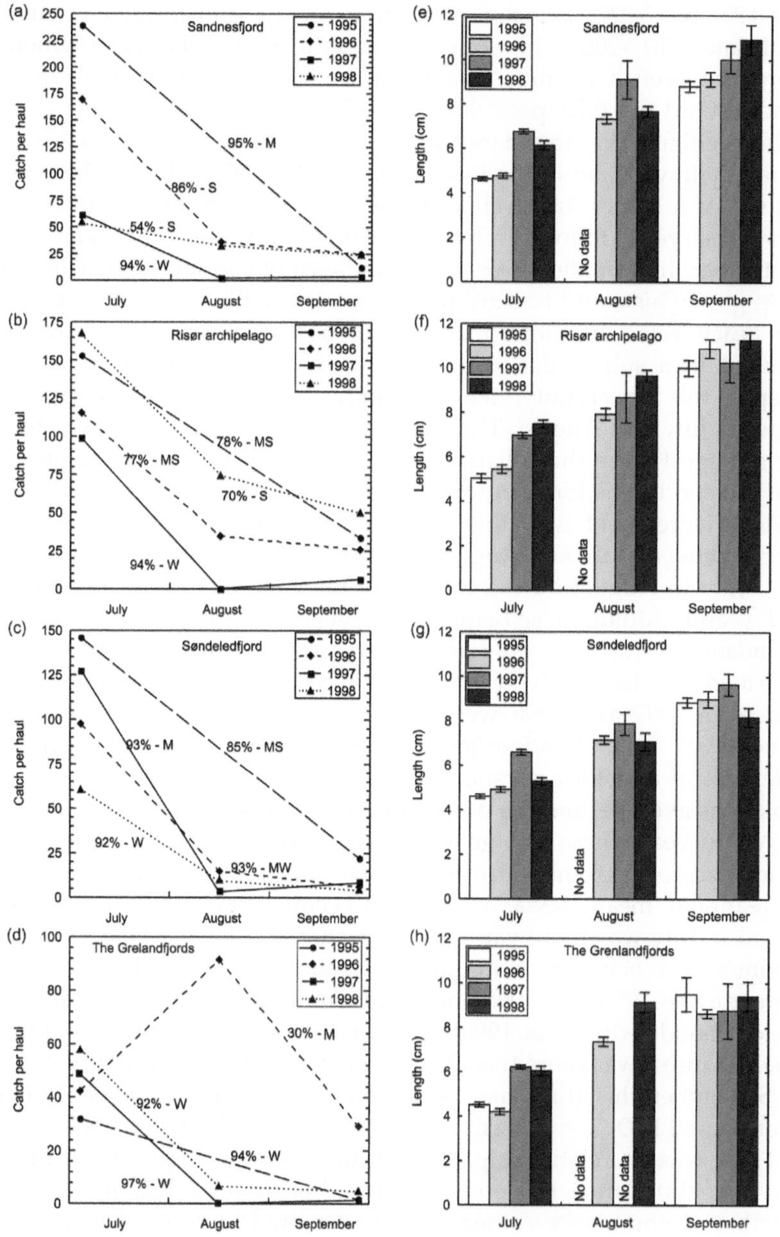

FIGURE 4.2 Average catch (left panels) and length (right panels) of 0-group cod in early July, mid-August, and late September 1985–1998 in various areas, except August 1995. Year-class strength is indicated as W–weak, M–medium, S–strong, or by combinations, evaluated on data 1981–2001, except in Grenland 1953–1965 (before the collapse). Vertical lines indicate 95% confidence interval for mean lengths. Notice the different scales.

($r = 0.22$, $p = 0.327$; Pearson correlation on log-transformed catch per haul data 1981–2001, the same period as used in Chapter 2), whereas there has been a positive correlation between Risør archipelago and Sandnesfjord ($r = 0.73$, $p < 0.001$, 1981–2001).

In Grenland the abundance of 0-group cod in July (Fig. 4.2d) was generally lower than in the Risør subareas. In 1997 and 1998 the typical pattern of a significant decline in the abundance from July to August was observed, followed by correspondingly low abundance in September. In 1996, however, the abundance of cod increased from July to August. This could be attributed to an unusually high catch at one of the beach seine locations (523 cod of a total of 825) that probably resulted in a substantial overestimation of the abundance in August. Despite this uncertainty, the data indicated no significant mortality between July and August. The year-class index in September was by far the strongest since the collapse in the mid-1960s (Fig. 2.7a), suggesting that the conditions for recruitment were unusually good in Grenland in 1996. In agreement with this, the average length of cod increased by 3.2 cm between July and August in 1996, which was the largest increase from July to August during this study (average 2.1 cm for all subareas and years). Although settlement was generally low in Grenland, the abundance in July was sufficient to give rise to strong year-classes, as indicated by the relatively strong 1996 year-class. In addition, abundances in 1997 and 1998 were higher than in 1996 and at about the same level as in Sandnesfjord in 1998. where a strong year-class emerged. In a historic perspective the 1996 year-class in Grenland was ranked as average, and the other three year-classes as weak.

Within years there was generally good correspondence between the mean sizes of 0-group cod from the various areas in July (Fig. 4.2e–h, intra-annual mean deviation $MD = 0.34$ cm). Between years, the size of cod in July was more variable (inter-annual $MD = 0.84$ cm, $p < 0.001$, Levine's test between intra- and inter-annual MD), with small fish appearing in 1995 and 1996 and relatively larger fish in 1997 and 1998 (except Søndeledfjord in 1998). The difference between the largest and smallest mean was more than 2 cm in all areas in July (average 2.14 cm). In September this difference was generally smaller (average 1.41 cm, inter-annual $MD_{July} = 0.84$ cm versus $MD_{Sept} = 0.50$ cm, $p = 0.007$).

There was no relationship between the average length of cod in July and the abundance in September ($r = -0.25$, $p = 0.362$, $n = 12$; both length and abundance estimates were centered to zero mean per subarea to avoid influences of local differences). In addition, the relative decrease (%) in abundance from July to September was independent of the length of cod in July ($r = 0.09$, $n = 12$, uncentered data). For example, the abundance of the large cod in July 1997 decreased by 94–97% in the various areas, resulting in weak year-classes in all areas except in

Søndeledfjord. The large cod in July 1998 resulted in strong year-classes in Sandnesfjord and Risør archipelago but a weak year-class in Grenland, and the small cod in July 1996 resulted in generally strong year-classes in all subareas except Søndeledfjord. Furthermore, there was no relationship between year-class strength and length in September ($r = 0.15$, $p = 0.589$; centered data), which is in agreement with analyses of the historical beach seine data (Fig. 3.4). However, there was a positive relationship between the length in July and that in September ($r = 0.55$, $p = 0.028$, $n = 12$; centered data), which suggests that the period prior to settlement has some impact on the size of 0-group cod as observed in September (Fig. 3.4).

4.3.2 Abundance of Other Littoral Fishes and Prawns

In July, fish of the same size as 0-group cod are potentially alternative prey for predators but also potential competitors for food. Most fish of similar size were I-group annual gobies (Chapter 3) that die shortly after spawning (Mesa et al., 2005, and references therein). In this study (Fig. 4.3), transparent goby (*Aphya minuta*) was the most

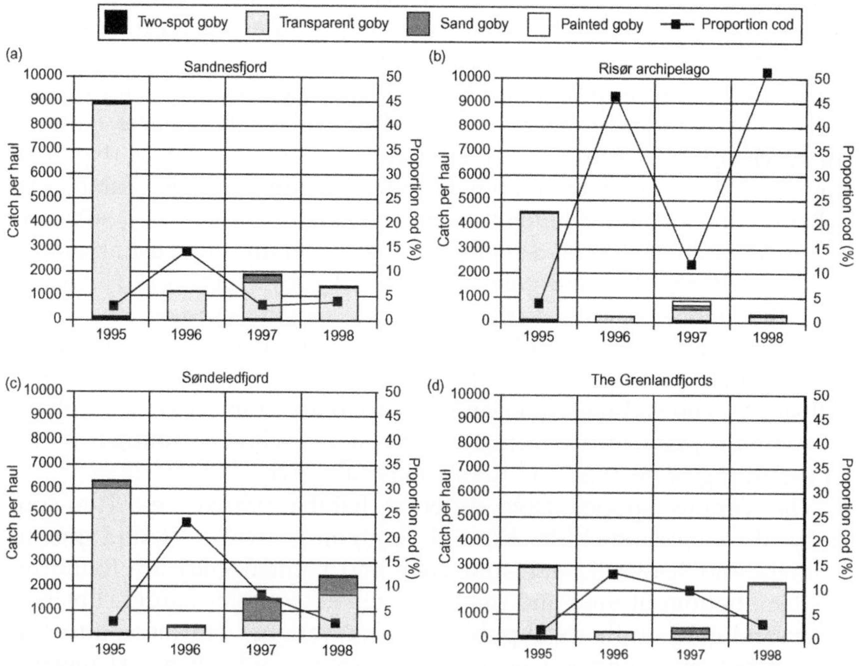

FIGURE 4.3 Average catch of similar size fish as settled cod July 1995–1999, and the proportion of cod in relation to the summarized abundance of these fishes.

abundant among fishes of similar fishes as settled cod: These pelagic fishes (Wheeler, 1969) probably gather in shallow water in the littoral zone to spawn in summer. The abundance varied substantially, from about 9000 individuals per haul in Sandnesfjord in 1995 to 200 per haul in Grenland in 1997. The second most abundant species, sand goby (*Pomatoschistus minutus*), showed also high intra- and inter-annual variability, from about 800 per haul in Søndeledfjord in 1997 and 1998 to 10 per haul in Risør archipelago and Grenland in 1996. The average catches of two-spot goby (*Gobiusculus flavenscens*) and painted goby (*Pomastoschistus pictus*) were generally less then 100 per haul.

The percentage of 0-group cod in relation to similar-sized fishes (gobies) varied substantially (Fig. 4.3), from 1.1% in Grenland in 1995 to 51.9% in Risør archipelago in 1998. There was no relationship between the abundance of similar-sized fish and percent reduction in abundance of 0-group cod from July to September ($\rho = 0.199$, $p = 0.442$, $n = 16$, Spearman Rank correlation).

In August, the abundance of littoral gobies and common prawn (*Palaemon serratus*) increased substantially by recruitment during summer (Fig. 4.4). These young-of-the-year organisms are substantially smaller than the 0-group cod, and at least the smallest stages are potential prey for cod. Young-of-the-year sand goby were often observed during the July surveys but could not be quantified because they were too small to be retained efficiently in the seine. As young-of-the-year transparent goby is pelagic in August, the beach seine catches do not reflect the abundance of this species. The overall abundances (sum of all species) varied substantially inter-annually in all subareas (Fig. 4.4). Grenland had the lowest abundances (2900 individuals, Table 4.1), whereas the overall abundances were similar in the Risør subareas (approximately 6000). Two-spot goby was generally the most abundant species, followed by sand goby and painted goby (Table 4.1). However, the abundance of the various species varied substantially inter-annually. In 1996 sand goby dominated the catches in all subareas except Risør archipelago, and two-spot goby dominated in 1997. All species showed high inter-annual variability (Table 4.1), particularly in common prawn for which the weakest year-class constituted 0–4% of the strongest year-class in the various subareas.

The various species have different habitat preferences. Two-spot goby, three-spine sticklebacks, and common prawn are found among eelgrass and seaweed, and sand goby and painted goby are found on the bare bottom of mud and sand (painted goby prefer harder substrate than sand goby; author's unpublished observations). However, observation of bottom vegetation using an aqua-scope did not reveal marked changes in bottom vegetation during this study, which was also the case generally along the Skagerrak coast during these years (Fig. 2.4f).

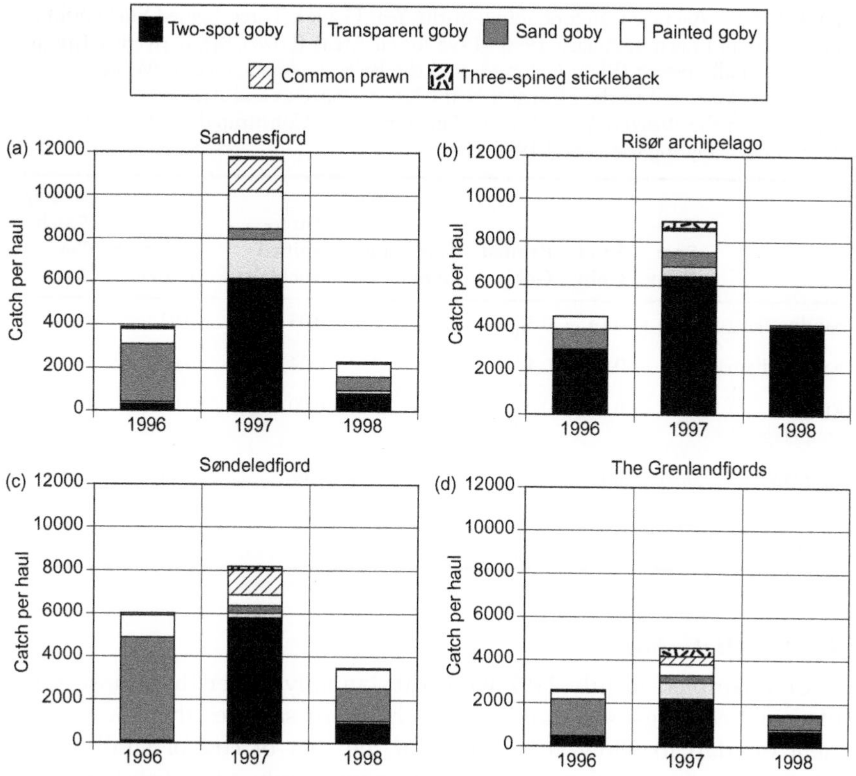

FIGURE 4.4 Average catch of young-of-the-year fish and common prawn in August 1996–1999 (no sampling in 1995).

Therefore, the observed variability in the abundance of non-gadoids (and cod) was probably not related to changes in habitat.

There was a tendency toward a larger reduction in the abundance of 0-group cod from July to September with higher abundance of young-of-the-year gobies and common prawn ($\rho = 0.538$, $p = 0.0745$, $n = 12$).

4.3.3 Diet and Condition in Settled Cod

Diet in settled cod was studied in Sandnesfjord, Risør archipelago, and in Grenland in 1996 and 1997, corresponding to periods when a strong (1996) and a weak (1997) year-class appeared in all areas. Data from Sandnesfjord and the Risør archipelago were pooled because both mortality patterns and diet were quite similar for the two areas during the study period (hereafter called Risør). The main focus is on July because this is probably the critical period for the recruitment of cod in these waters (Chapter 3).

TABLE 4.1 Average Catch of Young-of-the-Year Gobies, Three-Spined Stickleback, and Common Prawn in August 1996–1998 for all Areas (Lower Row), Average Catch per Area of all Species (Right Column), and the Relative Abundance of Weakest Year-Class Compared to the Strongest Year-Class per Species and Area (e.g., in Sandnesfjord the Weakest Year-Class of Common Prawn Constituted 3% [0.03] of the Strongest During the Period 1996–1998)

	Two-Spot Goby	Sand Goby	Painted Goby	Transparent Goby	Three-Spined Stickleback	Common Prawn	Average Catch Per Area
Sandnesfjord	0.06	0.19	0.37	0.04	0.03	0.03	5984
Søndeledfjord	0.01	0.07	0.48	0.14	0.24	0.00	5889
Risør arcipelago	0.47	0.07	0.09	0.06	0.00	0.03	5889
Grenland	0.20	0.19	0.06	0.06	0.02	0.04	2886
Average catch	2572	1223	678	318	94	277	

4.3.3.1 Diet—July

Cod sampled in July 1997 were substantially larger than those sampled in 1996 (Figs. 4.2 and 4.6). For the Risør samples there was sufficient overlap to allow comparison of the diet of cod of similar size, whereas for the Grenland samples there were only a few fish that overlapped (i.e., for subsamples used for stomach analyses, which were slightly different from the total samples). One location in Risør (location 95 in Sandnesfjord, Fig. 4.1) was excluded from the 1996 data because of a highly atypical diet dominated by a marine insect larva (see the following section).

In 1996 large copepods (P-pelagic) were the most important prey in Risør (Fig. 4.5a, total sample) by contributing to 42% of the volume (V, biomass) and 26% of the numbers (N), and appearing in 60% of the stomachs (F). The second most important prey was medium-sized copepods (V = 30%, F = 75%), but due to their relatively small size they were the most numerous prey (N = 63%). Fish (HB-hyperbenthic) contributed to 16% of the volume but were only present in 5% of the stomachs. Hence, pelagic prey dominated the diet in 1996 (V = 75%, N = 97%). Hyperbenthic prey (including benthic prey) were generally much larger than pelagic prey, as reflected in the relationship between volume and numbers (V = 25%, N = 3.2%, of which 2.2% were small mollusks).

In 1997 the importance of pelagic versus hyperbenthic prey was reversed as the two categories contributed to 25% and 75% of the

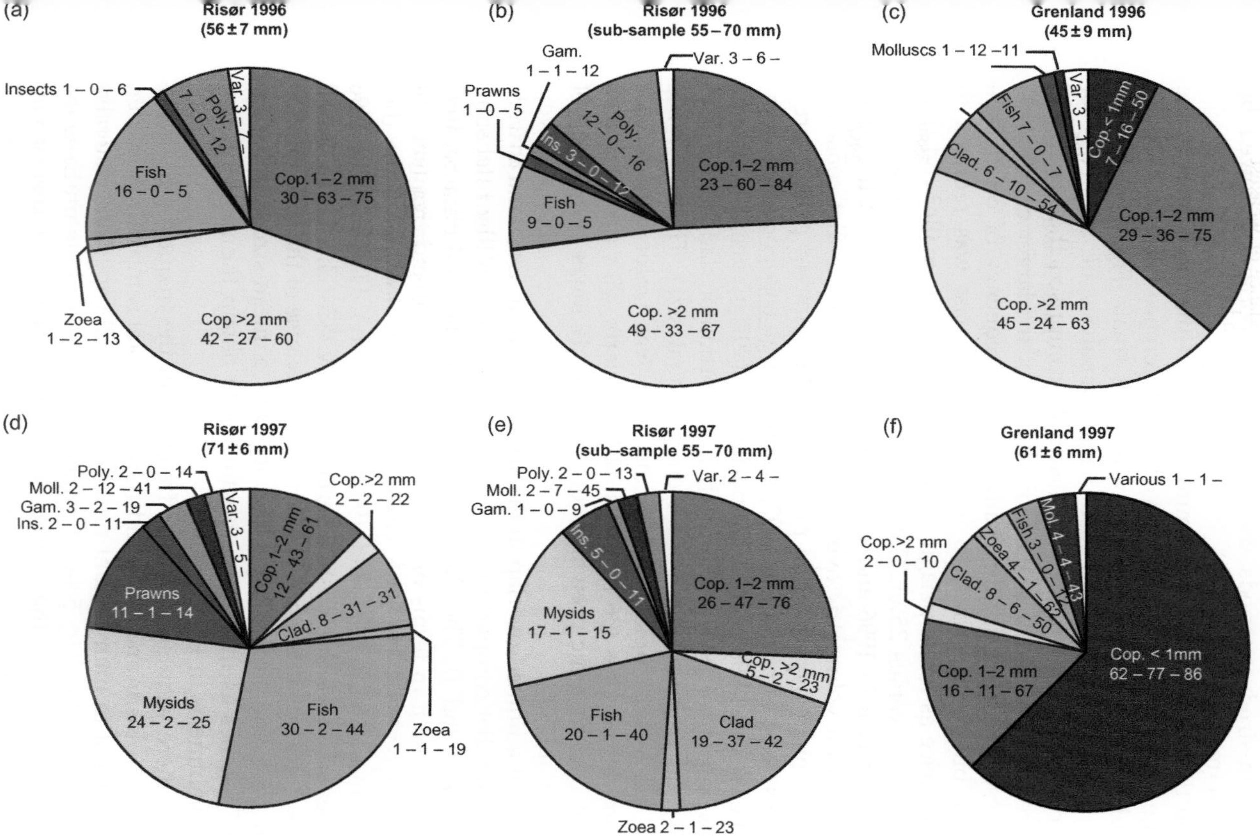

FIGURE 4.5 Diet in 0-group cod sampled in the Risør area and in the Grenland in July 1996 and 1997. Data from the Risør area are given for sub-sample consisting of cod 55–70 mm and the total samples. Numbers associated with prey: first number indicates percent of volume and corresponds to the cake piece, second number percent of numbers, and the third number is frequency of occurrence. Cop–copepods (P–pelagic), Clad–cladocerans (P), Zoea–*Eucharidae* larvae (P), Fish (HB–hyperbenthic), Mysids (HB), Prawns–*Caridea* (HB), Ins–insects (HP), Moll–molluscs (HP, benthic), Poly–polychaetes (HB, benthic), Var– various prey never occurring in more than 4% of volume in any sample (a–f).

volume, respectively (Fig. 4.5d, total sample). Fish, mysids, and shrimps were the main hyperbenthic prey ($V = 65\%$, $N = 4.6\%$). Among pelagic prey medium-sized copepods were the most important, followed by cladocerans (mainly *Podon* spp.).

In 1996 the subsample (55–70 mm) comprised the upper part of the size range of cod (Fig. 4.6a). However, the difference between the diet of the subsample and total sample was quite small (Fig. 4.5a, b; 85% overlap; $p = 0.889$, Chi-square contingency analysis of frequency of occurrence). In contrast, the 1997 subsample comprised the lower half of the size range of cod (Fig. 4.6c). Exclusion of the larger cod in 1997 did not result in a significant change in the frequency of occurrence ($p = 0.112$), but the diet overlap in terms of biomass was reduced to 69%, with twice as much pelagic prey in the group of smaller cod (Fig. 4.5d, e; $V = 51\%$ versus 25%). The diet of cod with similar size was significantly different in 1996 and 1997 (Fig. 4.5b, e; $p < 0.001$, Chi-square analysis), with a volumetric overlap of 42%. In 1997 the diet consisted of twice as much hyperbenthic prey as in 1996. Among pelagic prey, cladocerans were important in 1997, whereas large copepods that dominated the diet in 1996 were unimportant in 1997. Hence, the pelagic prey was smaller in 1997 with a higher proportion of cladocerans (cladocerans and medium copepods were of equal size). The difference between the total samples in 1996 and 1997 was larger than for the 55–70 mm subsamples (36% versus 42% overlap), with a higher proportion of hyperbenthic prey and correspondingly lower proportion of pelagic prey in 1997.

Because of a limited overlap in the size of cod from Grenland in 1996 and 1997, the diet comparison was made on the basis of the total samples (Fig. 4.5c and f). The diet of cod in Grenland in 1996 and 1997 was different both in terms of volume (36% overlap) and frequency of occurrence ($p < 0.001$). In both years pelagic prey dominated the diet ($V = 88\%$ and 92% in 1996 and 1997, respectively). However, in 1996 large copepods were the most important prey ($V = 54\%$), whereas small copepods dominated diet in 1997 ($V = 62\%$). Because of the combination of small prey and much larger fish in 1997, a mean-size cod had to capture 7.5 times as many prey organisms to obtain the same relative stomach volume as in 1996.

In 1996 there were significant differences in the frequency of occurrence between Risør and Grenland ($p < 0.001$, total samples). However, this was related to unimportant prey such as small copepods, cladocerans, and polychaetes (Fig. 5.4a, c), which did not contribute much to the volume ($p = 0.681$ with these prey groups excluded). In terms of volume, there was substantial overlap (81%) between the diet in Risør and Grenland, with large copepods being the most important prey, followed by medium copepods in both areas.

In 1997 the diet of cod from Risør differed substantially from cod sampled in Grenland, independent of whether the comparison was made on the basis of total sample (Fig. 4.5d, f, 28% overlap; $p < 0.001$, frequency of occurrence) or the subsample from Risør (Fig. 4.5e, 32% overlap).

4.3.3.2 *Condition—July*

Lipids are the major energy store in fish and in cod are mostly found in the liver (Black and Lowe, 1986). In small cod, the liver has been shown to rapidly change (within one to two weeks) with changes in food conditions (Grant and Brown, 1999). In 1996 the condition of the cod in Risør increased significantly with size (Fig. 4.6a; $p < 0.001$). A LOWESS smoothing (LOcally WEighted Scatter plot Smooth), which was carried out because of the pronounced nonlinear pattern in the data, indicated that the increase was steeper for the smaller cod, and that cod greater than 55 mm had approximately the same condition (the falling trend for the largest cod is due to one observation).

Because of accidental partial thawing of the July 1996 samples from Grenland, comparison with cod from Risør was done on the basis of the Fullton condition factor. Cod from Grenland had better condition than cod from Risør (Fig. 4.6b) with significant differences (<0.05, two-sided t-test) for all length intervals, except the smallest. In both areas the smaller cod had lower condition than larger cod. The indirectly estimated average liver index for Grenland was 4.2% versus 3.0% as measured in Risør. In 1997 the condition of cod in both Risør and Grenland decreased with size (Fig. 4.6c, d; $p = 0.013$ and $p = 0.010$, respectively).

The length interval 55–70 mm overlapped in all samples presented in Fig. 4.6 and corresponds approximately to the maximal liver index of the various samples. The liver index was significantly different between these subsamples (55–70 mm; $p < 0.001$ ANOVA; $p < 0.05$ Sheffe test between all sample pairs), with the highest liver index in Risør in 1996 (average 3.4%), followed by Grenland 1997 (3.0%), and Risør 1997 (2.5%). The average liver index in Grenland in 1996 for cod in the range of 55–70 mm was estimated to be 5.8%.

In addition to having the highest condition, cod in Grenland had significantly higher relative stomach content in 1996 (average $Si = 2.21\%$; $p < 0.001$ ANOVA) than in Risør 1996 (1.69%; $p = 0.002$ Sheffe test), Risør 1997 (1.62%; $p < 0.001$), and Grenland 1997 (1.42%; $p < 0.001$), whereas there were no significant differences between the three latter samples.

4.3.3.3 *Location 95—July 1996*

At location 95 in Sandnesfjord (Risør), the diet of cod in July 1996 was markedly different from the other locations in Risør, with larvae of the marine chironomid *Chironomus salinarius* dominating the diet

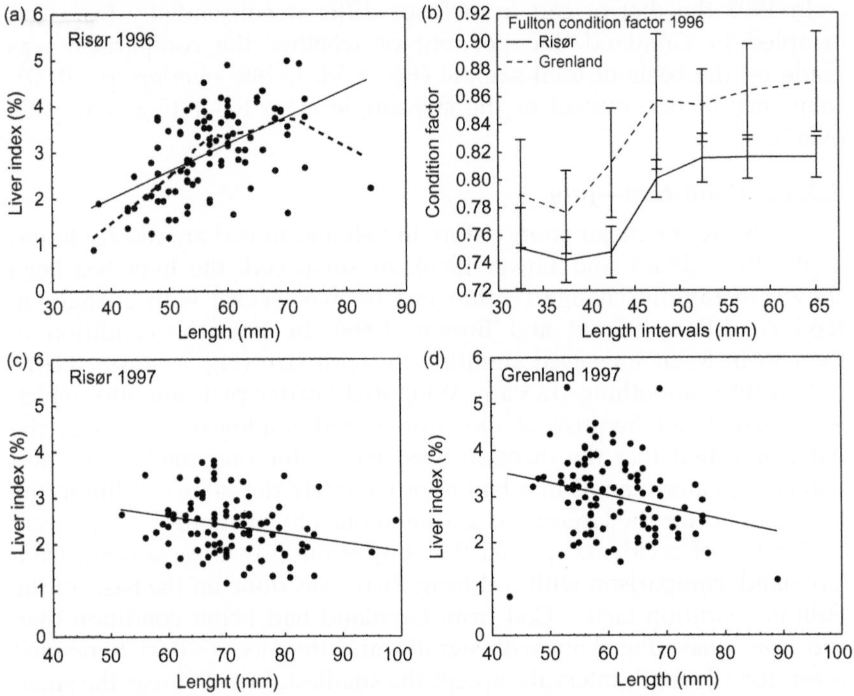

FIGURE 4.6 Panels a, c, and d: Liver index (weight of liver in percent of total body weight) versus length of 0-group cod caught in July 1996 and 1997 in the Risør area and Grenland (no data from the Grenland in 1996). The LOWESS smoothing in panel a was carried out using a tension factor of 0.25. Panel b: Fulton condition factor estimated for 5 mm intervals (except 60–69 mm) for cod caught in the Risør area and Grenland in July 1996 (fresh samples). Vertical lines indicate 95% confidence interval.

(V = 86%, N = 21%), followed by large copepods (V = 11%, N = 77%). The average liver index of these cod was 3.7%, indicating good condition. Cod probably preyed on the insect larvae as they emerged from bottom sediments to the sea surface. These insect larvae were also present in the diet of cod from this location in July 1997, although other prey dominated. The historical pattern in the abundance of gadoids at this location differed markedly from the other locations in this fjord (Fig. 2.5).

4.3.3.4 Diet and Condition—August and September

In August 1996 fish was the most important prey both in Risør and Grenland (Fig. 4.7a and c). In Risør, pelagic prey constituted close to 50% of the diet, with medium-sized copepods being the most important planktonic prey, followed by zoea larvae (approximately 2 times bigger than large copepods) and large copepods. In Grenland, 27% of the diet consisted of pelagic prey, with small copepods dominating this component of the diet. The overlap in diet between Risør and Grenland was

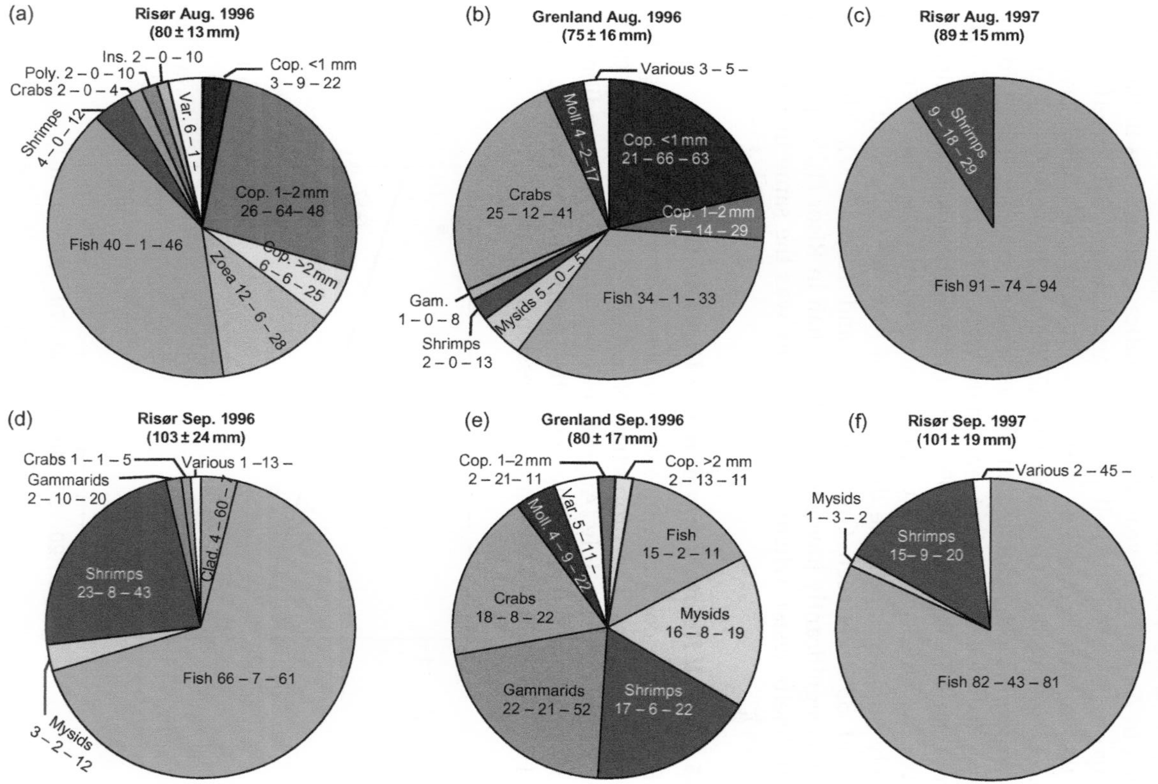

FIGURE 4.7 Diet in O-group cod sampled in the Risør area and in Grenland in August and September 1996 and in the Risør area in August and September 1997. Crabs–*Brachiura* (HB), Amphipods (HB). See Fig. 4.5 for explanation.

49% ($p < 0.001$, chi-square contingency analyses of F), with fish contributing the most of the overlap (34%). In August 1997, the diet of cod in Risør consisted exclusively of hyperbenthic prey with fish dominating (Fig. 4.7c, no data from Grenland in August 1997). The samples collected in August each year showed significant differences in the liver index ($p < 0.001$, ANOVA; all paired comparison $p < 0.05$, Sheffe test), with the highest condition observed in Risør in 1996 (average Li = 3.5%), followed by Grenland 1996 (2.2%) and Risør 1997 (1.7%).

Hyperbenthic prey dominated all September samples (Fig. 4.7d–f). In Risør, fish were the most important prey in both 1996 and 1997 (84% overlap). In Grenland (1996 data only), fish, mysids, amphipods, shrimps, and crabs were about equally important in the diet. There was relatively low overlap in the diet between Risør and Grenland (38% in 1996, and 33% between Grenland 1996 and Risør 1997). The liver index was significantly lower in Grenland than in Risør (1.7%, $p < 0.001$ for both Sheffe tests), where the liver index was the same in 1996 and 1997 (2.3%).

4.3.3.5 *Overall Relationship Between Diet and Condition*

In Fig. 4.8 the condition of cod in terms of average liver index has been plotted against the proportion of pelagic prey in the diet (total stomach samples from July, August, and September). The condition of the cod increased significantly with the proportion of pelagic prey ($a = 0.15$, $p = 0.003$, $r^2 = 0.68$, linear regression).

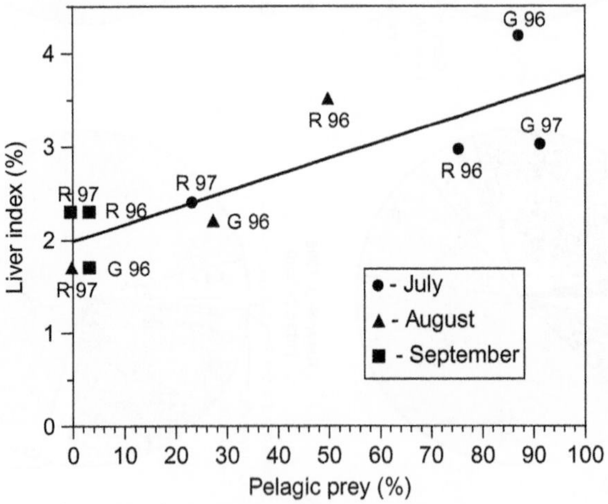

FIGURE 4.8 Liver index in cod in relation to the proportion of pelagic prey in the diet, sampled in July, August, and September in 1996 and 1997 in the Risør area (Sandnesfjord and Risør archipelago) and the Grenlandfjords. The liver index of cod from the Grenlandfjords in July 1996 was estimated indirectly.

4.4 DISCUSSION

4.4.1 Settlement, Mortality, Growth, and Year-Classs Strength in Cod

In July the abundance and size of 0-group cod varied substantially inter-annually (Fig. 4.2), but there was poor correspondence between the abundance in July and September. Mortality from July to September, as measured in terms of reduced abundance, was generally high (83%, mean of values in Fig. 4.2). Most of the mortality occurred within mid-August, which was reflected in low mortality rates between August and September and very good correspondence between the abundance estimates in these months. Furthermore, there was no relationship between the size of cod in July and mortality rates or between the size of cod in July and recruitment as measured in September. This suggests that larger cod did not have higher survival rates than smaller cod, which would probably have been the case if predation had been a major cause of mortality. The fact that similar size gobies in July generally outnumbered cod further suggests that predation was not an important cause of mortality.

On the other hand, fish of similar size could potentially compete with 0-group cod for food. However, there was no relationship between the abundance of similar size fishes and mortality in cod to support this hypothesis. The lack of competition may reflect low diet overlap or that the annual gobies do not feed during spawning.

There is good agreement between the results from this study and the Arendal study. The generally high mortality rates after settlement suggest that cod settled in surplus of the capacity of the benthic habitat, also in the Grenlandfjords, where recruitment collapsed in the mid-1960s. Based on analyses of the potential impact settlement above or below the capacity of the benthic habitat might have on growth and mortality in 0-group cod, it was suggested that settlement below the capacity of the benthic habitat is rare (Fig. 3.5c and related text). Hence, the results from both studies as well as the theoretical considerations are in agreement with prediction 1: The number of cod that settle does not limit recruitment; hence, recruitment is generally decoupled from the larval and post-larval phase.

Furthermore, both studies provided substantial evidence to suggest that predation was not the cause of mortality. Hence, we are left with food availability as the most likely cause of variable survival in settled cod, both between years and subareas.

4.4.2 Diet and Condition in Settled Cod

The condition of the cod increased significantly with the proportion of pelagic prey (Fig. 4.8). As the samples consist of increasingly larger fish from July to September, it could be argued that the decrease in condition was related to size rather than a higher proportion of hyperbenthic prey. However, there are several arguments against this. First, the condition of cod in both Risør and Grenland increased with size in July 1996 (Fig. 4.6). Second, in July 1996, cod in Risør with a high proportion of pelagic prey had significantly better condition than similar size cod (55–70 mm) in 1997 when the cod had consumed a higher proportion of hyperbenthic prey (Fig. 4.5). Relative stomach content in 1996 and 1997, however, was the same (Si = 1.67% and 1.62%, respectively). Third, in August 1996, cod in Risør with a relatively high proportion of pelagic prey had significantly better condition (Li = 3.5%) than both cod in Risør in August 1997 (Li = 1.7%) and in Grenland in August 1996 (Li = 2.2%) with high proportions of hyperbenthic prey (Fig. 4.7). One possible contributing factor to the low condition in cod in August 1997 could have been the high sea surface temperature. However, in Søndeledfjord a medium year-class emerged in 1997, suggesting that the high temperatures per se did not limit survival, which probably would have been the case if temperature had severely affected the condition of the cod. Hence, we conclude that the positive relationship between condition and proportion of pelagic prey is indeed related to prey type and not to the size of cod.

In August 1996, the abundance of young-of-the-year gobies and shrimps, which are potential prey for cod, was relatively high. Nevertheless, cod had consumed a relatively high proportion of pelagic prey and were accordingly in good condition. This supports the hypothesis that cod prefer pelagic prey well after settlement, and that the transition from pelagic to hyperbenthic prey is associated with a substantial drop in condition of the cod. Consequently, pelagic prey (mainly copepods) appears to have significantly higher nutritional value for newly settled cod than hyperbenthic prey. On the other hand, from August onwards cod appear to cope with the lower nutritional value of hyperbenthic prey, as reflected in lower and more stable mortality rates. Bigger cod may thus be able to prey efficiently on young-of-the-year hyperbenthic prey that becomes highly abundant from August onwards (Fig. 3.3 and Fig. 4.3).

Copepods constitute approximately 80% of mesozooplankton of the oceans (Verity and Smetacek, 1996). To cope with periods of low food availability, copepods in polar and temperate latitudes lay down reserve lipids that have high-energy storage capacity (39 kJ g^{-1}) relative to proteins and carbohydrates (17–18 kJ g^{-1}) (Lee et al., 2006). Copepods derive their lipids from phytoplankton, and evidence suggests that the highest level of wax esters (the main lipid group) is found in strictly herbivores zooplankton and decreases from herbivores through omnivores to carnivores (Sargent and Falk-Petersen, 1988).

Calanus finmarchicus, the dominant species among copepods larger than 2 mm in this study, store lipids in oil sacs, and lipids may comprise more than 50% of the dry weight (Sargent and Falk-Petersen, 1988). Grant and Brown (1999) observed a rapid increase in the liver index of 0-group cod in response to increased consumption of *C. finmarchicus* with a high incidence of oil sacs, and abruptly declining liver index when the lipid-rich prey were no longer consumed. The lower nutritional value of hyperbenthic prey is probably a result of the low-energy content of proteins when compared to lipids. Cladocerans (including *Podon* spp.), which were relatively important pelagic prey in Risør in July 1997, have lower energy content than copepods in these waters as reflected in low C:P ratios (Gismervik, 1997).

At location 95 in Sandnesfjord (Risør), chironomid larvae of *C. salinarius* dominated the diet of 0-group cod in July 1996. These cod were in good condition (Li = 3.7%). Along the coasts of southern Europe, this chironomid has been reported to contribute to the diet of shorebirds and fish (e.g., Drake and Arias, 1995). The species has been studied on the west coast of Norway (Koskinen, 1968) but to our knowledge has not been reported to be an important food source for fish in these waters. Koskinen (1968) found that in one year most of the chironomid larvae emerged in June, whereas in another year the larvae emerged in both June and July. Consequently, in some years chironomid emergence may overlap with the critical period for recruitment of cod. The finding that location 95 has shown a substantial increase in the abundances of 0-group cod and whiting over the last 25 years while the other locations in Sandnesfjord have experienced significant reductions (Fig. 2.5) suggests that larvae of *C. salinarius* may be an important food source at this location and thus provides a reasonable explanation for the deviating historical trends.

4.4.3 Recruitment in Relation to Diet and Condition

Grant and Brown (1999) compared the condition of early juvenile cod sampled in the field with the condition of cod of the same age dying from starvation in the laboratory and concluded that there was no evidence of 0-group cod survival being limited by availability of prey. However, from the mere lack of evidence of severe starvation it cannot be concluded that food availability does not limit survival (Bollens et al., 1992; McNamara and Houston, 1987). For example, malnourished and weak prey is more likely to be caught because less energy is required to capture them, and malnourished individuals probably take higher risks and spend more time searching for food. These factors will increase predation risk. As weak and undernourished individuals experiencing food limitation may well be consumed by predators before displaying severe symptoms of starvation, it may be argued that although predation may be the cause of death, food availability is the underlying mechanism causing variable recruitment.

The proposed hypothesis for cod postulates that both growth and survival are limited by food availability. The critical period for survival can be viewed as a funnel where the wide opening corresponds to the abundance of cod just after settlement (always in surplus) and the narrow tube the number of cod that can pass through it. The width of the tube varies inter-annually according to food conditions. Those that pass through the critical period experience food limitation and reduced growth. In years with ample food supply, a larger number of cod survive, and food availability for the individual cod is kept at the same level as in years with poor food supply. The growth is therefore not affected (see elaboration in Fig. 3.5 and related text). In agreement with this, there was no relationship between the year-class strength and size of cod in September (Fig. 3.4). Therefore, we next examine the relationship between diet and condition and the number survivors rather than growth.

In Grenland, the strongest year-class by far of cod since the collapse in the mid-1960s emerged in 1996. Settlement was relatively low as measured in July and the cod were small (Fig. 4.2d, h). However, mortality between July and August seemed to be insignificant and growth was the highest observed during this study. This concurred with very good condition (Fig. 4.6b) and high stomach content consisting of relatively large copepods (Fig. 4.5c). However, the exceptionally good conditions for recruitment of cod in Grenland in 1996 were brief. By August 1996 the condition of cod in Grenland was significantly lower than in Risør, and this became even more pronounced in September.

In Inner Oslofjord, where there was a recruitment collapse around 1930, there was a similar event in 1938 when an exceptionally strong year-class of cod emerged (see the more detailed description in Chapter 3). This concurred with high abundance of *C finmarchicus* that was advected into the fjord in the early summer of 1938 (Wiborg, 1940). *C. finmarchicus* is an abundant species over wide areas of the North Atlantic (Sundby, 2000). Kaartvedt (1993) distinguished between drifting and resident plankton and characterized *C. finmarchicus* as a drifting copepod. Accordingly, Bucklin et al. (2000) found no evidence of a significant genetic structure of fjord populations for *C. finmarchicus*. Hence, the extraordinary good recruitment in Grenland in 1996 and Inner Oslofjord in 1938 was probably a result of advection of copepods into the fjords and not from local production.

A strong year-class emerged in Risør in 1996, also corresponding to a high proportion of large copepods in the diet of newly settled cod.

However, due to high settlement, a substantial decrease in abundance (83%) of cod from July to September was observed (the funnel effect). In July the condition of cod increased with size (Fig. 4.6a). Cod smaller than 47 mm had low liver index (Li = 1.7%), and this size group comprised 41% of the population. It is therefore likely that the smaller individuals have suffered higher mortality than the larger individuals this year.

In both Risør and Grenland post-settled cod were considerably larger in July 1997 than in 1996 (2.3 and 2.8 times heavier in Risør and Grenland, respectively). In spite of this, mortality rates were higher in both areas in 1997. This concurred with lower condition in cod, and, in contrast to 1996, the liver index decreased with increasing size of cod in both Risør and Grenland. In Risør this was probably due to a higher proportion of hyperbenthic prey in the diet of larger cod, as reflected in the subsample (small cod) versus total sample (Fig. 4.5). In Grenland hyperbenthic prey were insignificant. Here the lower liver index in larger cod was probably caused by the small copepods (less than 1 mm) not providing satisfactory food for larger individuals. Grant et al. (1998) reported similar results and suggested that a lower liver index in larger cod was associated with increased metabolic expenditure when feeding on small prey items. In July 1997 the larger cod may have suffered higher mortality than smaller cod in Risør as well as in Grenland. In support of this argument, in July cod were significantly larger in 1997 than in 1996, but of similar sizes in September.

In a study from the North Sea, Beaugrand et al. (2003) found a close relationship between fluctuations in zooplankton and recruitment of Atlantic cod. A marked reduction in recruitment in the 1980s coincided with a decrease in the size of calanoid copepods by a factor of two. Interestingly, the reduction in copepod size was most pronounced in July. Rothschild (1998) concluded that large year-class of either *Calanus* or *Paracalanus*/*Pseudocalanus* appeared to be required to produce a large year-class of cod in the North Sea. In a comprehensive review of recruitment in Atlantic cod stocks with particular reference to *C. finmarchicus*, Sundby (2000) concluded that recruitment–temperature relationships of Atlantic cod were proxies for food availability during the early stages.

In summary, this study demonstrates a positive relationship between the condition of cod in July and number of survivors, and that condition, in turn, is related to the quality of food, with large copepods being particularly favorable. Hyperbenthic prey appear not to provide newly settled cod with sufficient energy. From mid-August, though, cod appear to cope with the lower nutritional value of hyperbenthic prey, probably because the cod have become sufficiently large to prey efficiently on young-of-the-year hyperbenthic prey, which at that time have become highly abundant. As carnivorous copepods accumulate less lipid than strictly herbivorous copepods (Sargent and

Falk-Petersen, 1988), carnivorous copepods will probably have lower nutritional value for small recruiting cod. The deviating diet in cod from location 95 (chironomid larvae) in combination with high liver index provides an explanation for the diverging trend at this location and supports the idea of food-limited survival in settled 0-group cod.

Consequently, the results of this study are in agreement with prediction 2: Inter and intra-annual variability in the number of cod surviving through the early juvenile critical period is related to food conditions in terms of quantity and quality of the prey. Furthermore, food availability appeared to limit survival of newly settled cod in Grenland too. This suggests that the recruitment collapse that occurred in this area in the mid-1960s was a result of abrupt changes in the plankton community that deprived young-of-the year gadoids of adequate prey.

4.4.4 Energy Flow Pattern

The results of this study and the Arendal study provide evidence that recruitment of cod in these waters is determined shortly after settlement, depending on the abundance of energy-rich planktonic prey. Although some of the results might reflect local phenomenon for the Skagerrak coast, there are lessons from these studies that should have more general relevance. First, cod seem to depend on high-quality prey until they are quite large (approximately 8 cm). Second, planktonic prey appear to have higher nutritional value for small cod than hyperbenthic prey. Third, the repeated incidents of recruitment collapse in relation to increasing nutrient loads suggest that the plankton community may shift abruptly from one state to another. Direct and indirect evidence of shifts in the plankton community are reviewed in Chapter 5, and in Chapter 6 a mechanistic explanation for resilience (sensu Holling, 1973) in the plankton community is proposed. Fourth, variability in survival rate of cod implies that the abundance of high-quality planktonic prey is highly variable. This last point is the essence of the proposed recruitment hypothesis for cod, which, in a more general form, states: *The survival of 0-group cod is limited by food availability, and recruitment variability results from differences in food supply arising from inter-annual variability in the energy flow pattern at low trophic levels of the marine pelagic food web.* Variable energy flow is linked to variability in the plankton community structure that generates variable energy flow patterns to higher trophic levels.

The vast majority of marine teleost fishes (Leis, 2007) and a high proportion of benthic invertebrates (Pechenik, 1999; Thorson, 1950) are planktonic during early life stages. All small organisms depending on the pelagic community food are likely to be affected by variable energy flow pattern, including cod during the pelagic stage, other spring-spawning gadoids, and summer-spawners. In support of this argument, cod settled in highly variable numbers (Fig. 4.2), recruitment in 0-group pollock and whiting collapsed simultaneously with that of cod in polluted areas like Grenland, and the abundance of summer-spawners varied considerably inter-annually (Fig. 3.3, Fig. 4.4, Table 4.1). In a seminal paper, Thorson (1950, p. 1) states: "In the invertebrate populations on the level sea bottom, large fluctuations in numbers from year to year indicate species with a long pelagic larval life, while a more or less constant occurrence indicates species with a very short pelagic life or a non-pelagic development."

The idea of food-limited survival in fish during early life stages is not new (e.g., Anderson, 1988; Cushing, 1990; Hjort, 1914). However, except for correlation studies, which have often been dubious (Bollens et al., 1992; Cushing, 1990), there are surprisingly few field studies to confirm proposed hypotheses. Here, the recruitment hypothesis for cod was proposed on the basis of historical data (correlations) in order to test the mechanism underlying the repeated incidents of abrupt and persistent recruitment collapses in 0-group gadoids. Field studies have provided substantial evidence in support of the hypothesis. In Chapter 8, which deals with variability enhancing and variability dampening mechanisms in marine ecosystems, the relationship with other hypotheses suggesting food-limited survival during early life stages in fish is discussed.

4.4.5 Historical and Distributional Differences Between the Gadoids

In a review of size-selective mortality in the juvenile stages of teleost fishes, Sogard (1997, p. 1129) states: "The size variability has generated the logical intuitive hypothesis that larger or faster growing members of a cohort gain a survival advantage over smaller conspecifics via enhanced resistance to starvation, decreased vulnerability to predators, and better tolerance of environmental extremes." However, the results of this study suggest that the perception that "bigger is better" does not apply to the critical recruitment phase of cod. Rather, there seems to be an optimal size between post-settled cod and available prey that maximizes survival. The slow growth in 0-group cod (Fig. 3.2b) can therefore be viewed as selective adaptation to the prevailing food conditions in these waters. Furthermore, growth rate as an adaptation to prey type can provide mechanistic explanations for differences between the gadoids with respect to geographical distribution and historical pattern in abundance and intra-specific genetic differentiation over short geographical distances.

There are substantial differences between the growth rates in 0-group pollock (*Pollachius pollachius*), whiting (*Merlangus merlangius*), and cod, 1.25 mm d^{-1}, 0.72 mm d^{-1}, and 0.55 mm d^{-1}, respectively (Chapter 3). During earlier periods when all three species were abundant in Sandnesfjord/Risør archipelago, the occurrence of the slow-growing cod was highest in the inner part of the fjord, whiting in the inner to mid-part of the fjord, whereas the fast-growing pollock was most abundant in the archipelago (Fig. 2.5). This may reflect adaptation to different prey. Before World War 2, the 0-group pollock were relatively abundant in Sandnesfjord, except for the innermost part. During the period 1955–1975, the abundance of pollock was low throughout the fjord but was still relatively high in the archipelago. In contrast, the abundance of whiting had increased in the mid part of the fjord (but keep in mind the deviating trend at location 95, 5.5 km inside the fjord). During the period 1981–2001 pollock had collapsed in the archipelago as well, however, without being replaced by cod or whiting. The latter suggests that the collapse in pollock was not caused by competition between the gadoids. Rather, the pattern of changing distribution in pollock indicates that inner fjords have never had adequate prey for 0-group pollock, and that such prey were gradually displaced to the outer part of the fjord before eventually almost disappearing in the archipelago as well. As recruitment in pollock was reduced by 96% from the 1920s to the period after the mid-1970s, this indicates that the abundance of larger, energy-rich planktonic prey that can give rapid growth had decreased substantially, in particular in summer as pollock spawn in May–June and settle from July onwards (Fig. 3.2). One candidate for such a prey is *C. finmarchicus*, which has decreased substantially in the North Sea and Skagerrak (Reid et al., 2003). There has been an increase in *C. helgolandicus*, but this species peaks in the autumn (Planque and Fromentin, 1996).

The noncommercial poor-cod (*Trisopterus minutus*) has also collapsed along the Skagerrak coast (Fig. 2.4). Like pollock, poor-cod probably spawn in early summer and settle from August onwards (Chapter 3.). In contrast, saithe (*Pollachius virens*) that settle from May onwards and have grown quite large already in June (Fig. 3.2) have shown no trend in recruitment over time in these waters (Fig. 2.4). This indicates that there have been substantial changes in the summer plankton. Beaugrand et al. (2003) have documented such a change in North Sea and Skagerrak, specifically toward smaller copepods in summer. In Chapter 7 a mechanistic explanation as to how increasing nutrient loads may induce changes toward smaller plankton in summer is proposed.

In agreement with the perception that planktonic prey contribute to faster growth in more exposed coastal area, the size of 0-group cod is smaller in the inner part of the semi-enclosed Søndeledfjord (Fig. 4.2) than in Risør archipelago (Dannevig, 1949). Recently, Knutsen et al. (2010) reported that the 0-group cod in the inner part of the Søndeledfjord and Risør archipelago are genetically different, the latter being similar to cod in the North Sea. Different spawning grounds can obviously result in genetic differentiation. However, the distance from the inner Søndeledfjord to Risør archipelago (approximately 10 km) is probably too short to prevent the spread of cod eggs and larvae from inner to outer fjord and vice versa, as the pelagic phase of cod in these waters lasts for about three months. Hence, there is probably a mechanism that prevents young-of-the-year cod from the two populations to grow up in the habitat of the other population. Adaptations to different planktonic prey types may be such a mechanism, with cod in Risør archipelago being adapted to larger oceanic planktonic prey and cod in inner Søndeledfjord to smaller estuarine zooplankton. In support of this hypothesis, recruitment in cod in Risør archipelago and Søndeldfjord is uncorrelated ($r = 0.22$), which suggests that the mechanisms underlying variability in the abundance of oceanic and estuarine plankton are different. Interestingly, cod recruitment in the more exposed Sandnesfjord is correlated with both Risør archipelago ($r = 0.73$) and Søndeledfjord ($r = 0.63$, $p = 0.002$), indicating both archipelago and fjord adapted populations. Adaptions to different prey may thus explain genetic differentiation over short distances and underlines the potential importance of growth rates during the critical recruitment phase as genetic adaptations to the prevailing food regimes.

4.5 CONCLUSION

A recruitment hypothesis for cod has been subjected to comprehensive testing in the field. The hypothesis predicts that food availability limits survival in settled young-of-the-year cod along the Norwegian Skagerrak coast, both in areas with and without recruitment collapses. Substantial evidence in support of the hypothesis has been provided. High numbers of survivors and good condition were observed with a diet of relatively large copepods, whereas low survival was concurred with lower condition and a diet of hyperbenthic prey like fish and prawns or small copepods. Hyperbenthic prey appeared not to provide sufficient energy for newly settled cod. Evidence suggests that the perception that "bigger is better" does not apply to the critical period for survival in cod. Rather, there seems to be an optimal size relationship between cod and available prey that maximizes survival. A consequence of these findings is that variability in the plankton community structure generates variable energy flow patterns to higher trophic levels and thereby induces recruitment fluctuations in fish and other organisms that depend on planktonic prey during early life stages. Furthermore, the repeated incidents of abrupt and persistent recruitment collapses in gadoids in relation to increasing eutrophication appear to have been results of sudden changes in the plankton communities.

References

Anderson, J.T., 1988. A review of size dependent survival during pre-recruit stages of fishes in relation to recruitment. J. Northw. Atl. Fish. Sci. 8, 55–66.

Beaugrand, G., Brander, K.M., Lindley, J.A., Souissi, S., Reid, P.C., 2003. Plankton effect on cod recruitment in the North Sea. Nature. 426, 661–664.

Black, D., Lowe, R.M., 1986. The sequential mobilisation and restoration of energy reserves in tissues of Atlantic cod during starvation and refeeding. J. Comp. Physiol. B. 156, 469–479.

Bollens, S.M., Frost, B.W., Schwaninger, H.R., Davis, C.S., Way, K.J., Landsteiner, M.C., 1992. Seasonal plankton cycles in a temperate fjord and comments on the match-mismatch hypothesis. J. Plankton Res. 14, 1279–1305.

Bucklin, A., Kaartvedt, S., Guarnieri, M., Goswami, U., 2000. Population genetics of drifting (*Calanus spp.*) and resident (*Acartia clausi*) plankton in Norwegian fjords. J. Plankton Res. 22, 1237–1251.

Burrow, J.F., Horwood, J.W., Pitchford, J.W., 2011. The importance of variable timing and abundance of prey for fish larval recruitment. J. Plankton Res. 33, 1153–1162.

Cushing, D.H., 1990. Plankton production and year-class strength in fish populations: an update of the match/mismatch hypothesis. Adv. Mar. Biol. 26, 249–293.

Dannevig, A., 1949. The variation in growth of young codfishes from the Norwegian Skagerrak coast. Fiskeridir. Skr. Ser. Havunders. 9, 1–12.

Drake, P., Arias, A.M., 1995. Distribution and production of *Chrionomus* salinarius (Diptera: Chironomidae) in a shallow coastal lagoon in the Bay of Cádiz. Hydrobiologia. 299, 195–206.

Fromentin, J.M., Stenseth, N.C., Gjosaeter, J., Johannessen, T., Planque, B., 1998. Long-term fluctuations in cod and pollock along the Norwegian Skagerrak coast. Mar. Ecol. Prog. Ser. 162, 265–278.

Gibson, R.N., Ezzi, I.A., 1987. Feeding relationships of a demersal fish assemblage on the west coast of Scotland. J. Fish Biol. 31, 55–69.

Gismervik, I., 1997. Stoichiometry of some marine planktonic crustaceans. J. Plankton Res. 19, 279–285.

Gotceitas, V., Fraser, S., Brown, J.A., 1997. Use of eelgrass beds (*Zostera marina*) by juvenile Atlantic cod (*Gadus morhua*). Can. J. Fish. Aquat. Sci. 54, 1306–1319.

Grant, S.M., Brown, J.A., 1998a. Nearshore settlement and localized populations of age 0 Atlantic cod (*Gadus morhua*) in shallow coastal waters of Newfoundland. Can. J. Fish. Aquat. Sci. 55, 1317–1327.

Grant, S.M., Brown, J.A., 1998b. Diel foraging cycles and interactions among juvenile Atlantic cod (*Gadus morhua*) at a nearshore site in Newfoundland. Can. J. Fish. Aquat. Sci. 55, 1307–1316.

Grant, S.M., Brown, J.A., 1999. Variation in condition of coastal Newfoundland 0-group Atlantic cod (*Gadus morhua*): field and laboratory studies using simple condition indices. Mar. Biol. 133, 611–620.

Grant, S.M., Brown, J.A., Boyce, D.L., 1998. Enlarged fatty livers of small juvenile cod (*Gadus morhua*): a comparison of laboratory cultured and wild juveniles. J. Fish Biol. 52, 1105–1114.

Heath, M.R., Lough, R.G., 2007. A synthesis of large-scale patterns in the planktonic prey of larval and juvenile cod (*Gadus morhua*). Fish. Oceanogr. 16, 169–185.

Hjort, J., 1914. Fluctuations in the great fisheries of northern Europe viewed in the light of biological research. Rapp. P.-v. Réun. Cons. int. Explor. Mer 20, 1–228.

Holling, C.S., 1973. Resilience and stability of ecological systems. Annu. Rev. Ecol. Syst. 4, 385–398.

Houde, E.D., 2008. Emerging from Hjort's shadow. J. Northw. Atl. Fish. Sci. 41, 53–70.

Johannessen, T., Tveite, S., 1989. Influence of various physical environmental factors on 0-group cod recruitment as modelled by partial least-square regression. Rapp. P.-v. Réun. Cons. int. Explor. Mer 191, 311–318.

Kaartvedt, S., 1993. Drifting and resident plankton. B. Mar. Sci. 53, 154–159.

Knutsen, H., Olsen, E.M., Jorde, P.E., Espeland, S.H., André, C., Stenseth, N.C., 2010. Are low but statistically significant levels of genetic differentiation in marine fishes "biologically meaningful?" A case study of coastal Atlantic cod. Mol. Ecol. 20, 768–783.

Koskinen, R., 1968. Seasonal and diel emergence of *Chrionomus salinarius* Kieff. (Dipt., Chironomidae) near Bergen, Western Norway. Ann. Zool. Fenn. 5, 65–70.

Kristiansen, T., Drinkwater, K.F., Lough, R.G., Sundby, S., 2011. Recruitment variability in North Atlantic cod and match-mismatch dynamics. PLoS ONE 6, e17456.

Lee, R.F., Hagen, W., Kattner, G., 2006. Lipid storage in marine zooplankton. Mar. Ecol. Prog. Ser. 307, 273–306.

Leis, J.M., 2007. Behaviour as input for modelling dispersal of fish larvae: behaviour, biogeography, hydrodynamics, ontogeny, physiology and phylogeny meet hydrography. Mar. Ecol. Prog. Ser. 347, 185–193.

McNamara, J.M., Houston, A.I., 1987. Starvation and predation as factors limiting population size. Ecology 68, 1515–1519.

Mesa, M.L., Arneri, E., Caputo, V., Iglesias, M., 2005. The transparent goby, *Aphia minuta*: review of biology and fisheries of a paedomorphic European fish. Rev. Fish. Biol. Fisher. 15, 89–109.

Methven, D.A., Bajdik, C., 1994. Temporal variation in size and abundance of juvenile Atlantic cod (*Gadus morhua*) in an inshore site off eastern Newfoundland. Can. J. Fish. Aquat. Sci. 51, 78–90.

Pechenik, J.A., 1999. On the advantages and disadvantages of larval stages in benthic marine invertebrate life cycles. Mar. Ecol. Prog. Ser. 177, 269–297.

Planque, B., Fromentin, J.-M., 1996. Calanus and environment in the eastern North Atlantic. 1. Spatial and temporal patterns of *C. finmarchicus* and *C. helgolandicus*. Mar. Ecol. Prog. Ser. 134, 101–109.

Reid, P.C., Edwards, M., Beaugrand, G., Skogen, M., Stevens, D., 2003. Periodic changes in the zooplankton of the North Sea during the twentieth century linked to oceanic inflow. Fish. Oceanogr. 12, 260–269.

Rothschild, B.J., 1998. Year class strength of zooplankton in the North Sea and their relation to cod and herring abundance. J. Plankton Res. 20, 1721–1741.

Sargent, J.R., Falk-Petersen, S., 1988. The lipid biochemistry of calanoid copepods. Hydrobiologia 167/168, 101–114.

Smith, T.D., 1994. Scaling Fisheries. The Science of Measuring the Effect of Fishing. 1855–1955. Cambridge University Press, Cambridge.

Sogard, S.M., 1997. Size-selective mortality in Juvenile stages of teleost fishes: a review. Bull. Mar. Sci. 60, 1129–1157.

Sundby, S., 2000. Recruitment of Atlantic cod stocks in relation to temperature and advection of copepod populations. Sarsia 85, 277–298.

Thorson, G., 1950. Reproductive and larval ecology of marine bottom invertebrates. Biol. Rev. 25, 1–45.

Tupper, M., Boutilier, R.G., 1995a. Size and priority at settlement determine growth and competitive success of newly settled Atlantic cod. Mar. Ecol. Prog. Ser. 118, 295–300.

Tupper, M., Boutilier, R.G., 1995b. Effects of habitat on settlement, growth, and postsettlement survival of Atlantic cod (*Gadus morhua*). Can. J. Fish. Aquat. Sci. 52, 1834–1841.

Tveite, S., 1971. Fluctuations in year-class strength of cod and pollock in southeastern Norwegian coastal waters during 1920–1969. Fiskeridir. Skr. Ser. Havunders. 16, 65–76.

Verity, P.G., Smetacek, V., 1996. Organism life cycles, predation, and the structure of marine pelagic ecosystems. Mar. Ecol. Prog. Ser. 130, 277–293.

Wheeler, A., 1969. The Fishes of the British Isles and North-west Europe. MacMillian, London.

Wiborg, K.F., 1940. The production of zooplankton in the Oslofjord 1933–1934. Hvalråd. Skr. Sci. Results Mar. Biol. Res. 21, 1–87.

CHAPTER

5

Bifurcations in Marine Ecosystems: Concurrent Recruitment Collapses in Gadoid Fishes and Changes in the Plankton Community

5.1 INTRODUCTION

5.1.1 Ecosystem Bifurcations and Resilience

Most marine organic matter is produced by planktonic autotrophs (Charpy-Roubaud and Sournia, 1990) and the vast majority of marine teleost fishes (Leis, 2007), and a high proportion of benthic invertebrates (Pechenik, 1999; Thorson, 1950) depend on planktonic prey during early life stages. Understanding the dynamics of plankton communities is a prerequisite for analyzing drivers and processes in marine ecosystems. At present, we generally base our assessment and monitoring of dynamics of ecosystems on the assumption of simple dose-response relationships. Gradual environmental changes or perturbations are expected to cause corresponding changes in the abundance of affected species. However, it has long been recognized theoretically that ecosystems may shift between alternative stable states, each of which has its own basin of attraction (Holling, 1973; Lewontin, 1969; May, 1977). Given the prospect of global warming, a significant topical question is how marine plankton communities will respond to gradual environmental changes.

From an Antagonistic to a Synergistic Predator Prey Perspective.
DOI: http://dx.doi.org/10.1016/B978-0-12-417016-2.00005-1

In the marine literature the term regime shift has been frequently used to describe abrupt changes in time series. However, different definitions of regime shifts (Jarre et al., 2006; Overland et al., 2008) have rendered the concept vague. Therefore to be more explicit, the concept of ecosystem bifurcations is used here, referring to abrupt and persistent ecosystem shifts that affect several trophic levels and result from gradual environmental changes.

Resilience is a concept inseparably linked to ecosystem bifurcations. However, there are different definitions of the concept (Gunderson, 2000; Pimm, 1991). Here, resilience is used as proposed by Holling (1973): the maximum perturbation a system can sustain without causing a shift to an alternative stable state. The theoretical relationship between bifurcations, resilience and environmental perturbations is illustrated in Fig. 1.1.

Scheffer et al. (2001) reviewed evidence of shifts between contrasting states in large-scale ecosystems. Most of these examples were ecosystem shifts attributed to abrupt environmental shifts or catastrophic events (e.g., storms, mass mortality due to pathogens). One example, however, was the gradually increasing eutrophication in shallow lakes causing shifts from a clear water state with submerged vegetation to a turbid state in which phytoplankton dominated. In the clear water state, submerged vegetation held a key position that at a certain level of eutrophication shifted to a turbid state dominated by phytoplankton. In general, in ecosystems with one or a few keystone species (sensu Paine, 1969) shifts will occur when the tolerance limit for the keystone species is reached. However, in many ecosystems, including the diverse marine plankton community (Hutchinson, 1961), there are no obvious keystone species. Therefore, the question of shifts between contrasting states still remains controversial, and abrupt ecosystem shifts are counterintuitive to many (Scheffer and Carpenter, 2003), in particular the idea of bifurcations. One reason for this scepticism is probably because there are few repeated incidents from real ecosystems to confirm that observed shifts are results of gradual environmental changes.

5.1.2 Empirical Background

Along the Norwegian Skagerrak coast, repeated incidents of abrupt and persistent recruitment collapses in gadoids have been observed and related to increasing eutrophication (Chapter 2). Comprehensive testing in the field provided substantial evidence to suggest that the recruitment collapses were caused by abrupt shifts in the plankton community that deprived the young-of-the-year gadoids of adequate prey (Chapters 3 and 4). Here, various published time series have been updated that provide direct and indirect evidence of abrupt changes in the plankton community, for example, data on oxygen (Johannessen and Dahl, 1996a, 1996b), gadoid abundance (Chapter 2), and phytoplankton and zooplankton (Johannessen et al., 2012). In addition, a more recent incident of concurrent recruitment failure in gadoids and changes in plankton community is presented (Johannessen et al., 2012). In contrast to the previous events, this recent shift occurred during increasing temperature and decreasing eutrophication. This contrasts with shifts observed and analyzed previously, which is the reason why this shift has been treated separately.

5.2 METHODS

As most of the time series presented in this chapter are updates of previously published data, only basic information about the methods are presented here.

5.2.1 Beach Seine Sampling

A beach seine sampling program has been carried out annually since 1919 (except 1940–1944) along the Norwegian Skagerrak coast (details in Chapter 2) with no methodological changes since the start of the program. In this chapter, trends in the abundance of young-of-the-year gadoids, Atlantic cod (*Gadus morhua*), pollack (*Pollachius pollachius*), and whiting (*Merlangius merlangus*), sampled at 38 fixed locations (between Torvefjord and Kragerø, Fig. 2.1) are presented (updates of Fig. 2.3a–c). In addition, the abundance of ≥ I-group cod and the sum of older gadoids, cod, pollack, and saithe (*Pollachius virens*) sampled at the same 38 locations is presented. The time series were smoothed by computing the seven-year moving average twice (i.e., smoothing the raw time series and then smoothing the smoothed time series; Chapter 2).

Jellyfish and ctenophores are important consumers of fish larvae and zooplankton and hence constitute both predators and potential competitors of fish (Purcell, 2005). Ctenophores occurred in beach seine catches but were not quantified. The abundance has generally been low, and there have been no conspicuous changes over time. The number of jellyfish of the species *Cyanea* cf. *capillata* (some *C. lamarckii*) caught in the beach seine has been counted, whereas the abundance of common jellyfish, *Aurelia aurita*, was recorded semiquantitatively on a 5-point scale: 1–one medusa, 2–a few, 3–some, 4–many, and 5–numerous medusa. The annual abundance of *C. capillata* was estimated in terms of number caught per beach seine haul, and common jellyfish was estimated as the average of the 5-point code records. The time series were smoothed by computing the seven-year moving average twice.

In response to the extensive mortality of eelgrass (*Zostera marina*) throughout the North Atlantic in the early 1930s (Short et al., 1988), semiquantitative observations of bottom vegetation (mainly eelgrass, but also some macroalgae) at the beach seine locations were recorded since 1934. From these observations, an index of vegetation coverage was estimated on a relatively coarse scale. Due to poor visibility, precision of the records was rather low in some years; hence only the trend in terms of the double seven-year moving average, as described previously, is presented here (but see Fig. 2.4f).

5.2.2 Phytoplankton

The concentration of Chl *a* at 0–3 m depth was measured three times a week in the Flødevigen Bay (Fig. 3.1) using the standard fluorescent method of Strickland and Parsons (1968). Trends in Chl *a* between 1989 and 2012 were studied on the basis of annual averages and separately for spring (Julian day 20–119), summer (day 120–214), and autumn (day 215–314; for details see Johannessen et al., 2012). The annual pattern in Chl *a* was estimated for two periods, 1989–2001 and 2002–2012. The choice of 2002 as the start of the second period was based on a significant change in the autumn bloom conditions that year (Fig. 5.2c). For each period the data were pooled according to Julian days (1–360) and then smoothed by estimating the 15 d moving average twice.

5.2.3 Copepods

Zooplankton sampling was carried out fortnightly between 2000 and 2012. The sampling station is situated near Flødevigen Bay (Fig. 3.1), approximately 1 nautical mile offshore in the coastal current. The water depth at the site is 105 m and vertical tows were made from 50 m to the surface using a WP2 plankton net ($0.25\ m^2$, 180 μm mesh size; UNESCO, 1968). Each sample was split using a Motoda splitter device (Motoda, 1959). Half of the sample was preserved in 4% borax-buffered formaldehyde-seawater solution for species identification and enumeration. (The other half was used to derive biomass estimates not used in this study.) Variation in abundance of copepods is presented for each of the following species or genera: *Paracalanus/Pseudocalanus*, *Calanus* spp. (*C. finmarchicus* and *C. helgolandicus*), *Acartia* spp., *Temora longicornis*, *Centropages* spp., *Metridia* spp., and *Oithona* spp.

5.2.4 Oxygen

Since 1927, oxygen concentration was measured at 31 fixed stations along the Skagerrak coast. The measurements were obtained during the beach seine survey using standard Winkler procedure (Strickland and Parsons, 1968) throughout the sampling period. At all analyzed depths, ≥ 10 m oxygen concentrations decreased during the period 1927–1993 (Johannessen and Dahl, 1996a). Here, overall trends (all stations) in oxygen at 30 m depth and bottom water (variable depths) have been updated until 2012. Prior to estimating an annual average for all stations (except stations with permanent anoxia), oxygen saturation was standardized for all stations (zero mean and unit variance), which imply that all stations have the same impact on the estimated trend. Standardized values are in relative units (standard deviations). The real oxygen saturation per station and the position of the stations can be seen in Johannessen and Dahl (1996a). The time series were smoothed by computing the seven-year moving average twice.

5.2.5 Temperature

Temperature has been measured daily at 1 m depth since 1924 in Flødevigen Bay. The lowest sea surface temperature in these waters generally occurs in February and March and the highest temperature in July and August. Based on these daily measurements, trends in average annual, winter, and summer temperatures have been estimated.

5.3 RESULTS

5.3.1 Beach Seine Data

Because several of the time series presented here are updates of time series presented in Fig. 2.4, the main focus will be on the updated period, after 2001. The abundance of 0-group cod and pollack decreased substantially in the early 1930s (Fig. 5.1a, b), concurrent with a disease that wiped out eelgrass along the coasts of the North Atlantic (Short et al., 1988). Parallel with the recovery of bottom coverage in the 1950s, the abundance of cod, pollack, and whiting increased. Pollack never regained its former abundance though, whereas whiting increased substantially above the abundance in the 1920s. However, because of limited spatial overlap between the two species a direct causal relationship through competition is unlikely (Chapter 2). In the mid-1970s the abundance of whiting dropped abruptly. There was also a substantial decrease in the abundance of cod, and 1976 was the last

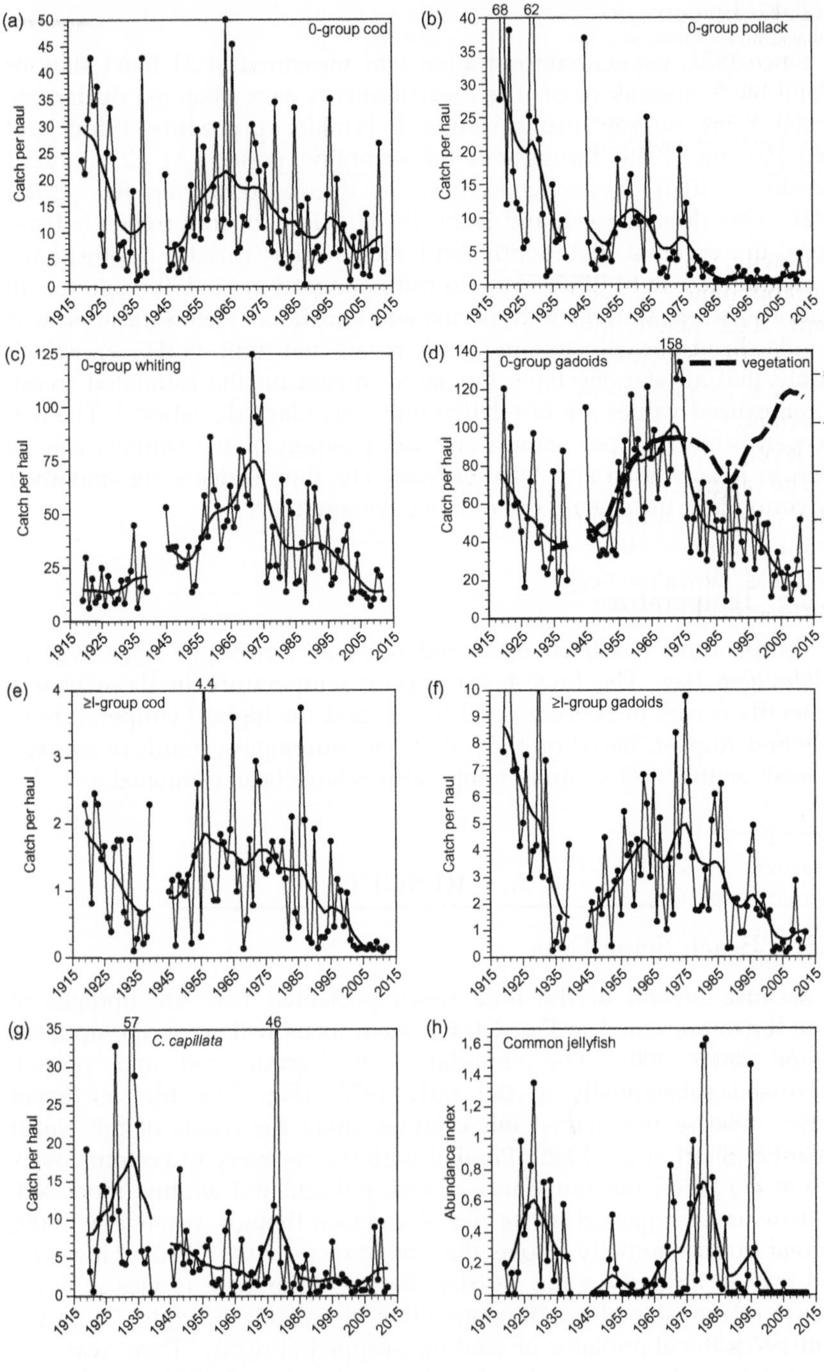
(a)
0-group cod
Catch per haul
(b)
0-group pollack
68
62
(c)
0-group whiting
(d)
0-group gadoids
158
vegetation
Coverage (%)
(e)
≥I-group cod
4.4
(f)
≥I-group gadoids
(g)
C. capillata
57
46
(h)
Common jellyfish
Abundance index
1915 1925 1935 1945 1955 1965 1975 1985 1995 2005 2015

FIGURE 5.1 Average catch of 0-group (a) cod; (b) pollack; (c) whiting; (d) sum of cod, pollack, and whiting and bottom flora coverage (thick line); (e) ≥ I-group cod; (f) sum of ≥ I-group cod, pollack, and saithe; (g) *C. capillata*; and (h) common jellyfish at 38 beach seine location along the Skagerrak coast 1919–2012. Smoothed curves correspond to seven-year moving average estimated twice (see text for details). For common jellyfish, an annual index of abundance was estimated as the average of semiquantitative abundance codes (see text).

year with reasonably high abundance in pollack. After the mid-1970s the abundance of whiting fluctuated around a relatively stable level until 2002 when another abrupt drop in the abundance occurred ($p < 0.001$, STARS, cut-off length of 10, target probability of 0.05, Huber's weight parameter set to 5 as in Chapter 2; Rodionov and Overland, 2005). After 2000 the abundance of cod was also at the lowest level on record.

The aggregate abundance of 0-group cod, pollack, and whiting showed a development corresponding with the pattern in bottom vegetation coverage with a substantial decrease in the abundance in 1930s (eelgrass mortality occurred in 1933 in these waters) followed by an increase in the 1950s. The abrupt drop in the gadoid abundance in the mid-1970s was decoupled from variation in vegetation (Fig. 5.1d). Since 2000, bottom vegetation coverage has been the highest on record, whereas the recruitment of gadoids has been the lowest.

The abundance of older cod (mainly I-group) followed the pattern in abundance in 0-group until about 2000 (Fig. 5.1e). Accordingly, a positive relationship was found between the abundance of 0-group and I-group of the same year-class between 1919 and 2001 ($r^2 = 0.72$, Chapter 2). However, after 2000 the beach seine catches were very low and correspondence poor between the year-class abundance at 0-group and I-group stage. One example is the very low catches of older cod in 2012 following what appeared as an abundant 2011 year-class measured at the 0-group stage.

In the beach seine catches older gadoids constituted 80% of predominantly piscivorous fishes ≥20 cm between 1919 and 2001 (Chapter 2), of which cod, pollack, and saithe contributed 42%, 49%, and 9%, respectively. After 2001 the abundance of older gadoids was the lowest on record (Fig. 5.1f).

The abundance of the jellyfish *C. capillata* apparently decreased substantially since the 1920s ($p < 0.001$, linear regression on log-transformed annual averages, Fig. 5.1g). Common jellyfish showed no clear trend but had peak abundances around 1930 and 1980 (Fig. 5.1h). The lowest abundance on record was observed between 1998 and 2012, when only two common jellyfish were caught.

5.3.2 Phytoplankton

There have been decreasing trends in Chl *a* (Fig. 5.2) in summer ($p = 0.019$, linear regression) and autumn Chl *a* ($p < 0.001$, linear regression on log-transformed Chl *a* to stabilize variance), but no trend in spring ($p = 0.410$). In autumn 2002 an abrupt shift occurred ($p < 0.001$, STARS). The annual patterns in Chl *a* during the periods 1989–2001 and 2002–2012 revealed extensive autumn blooms during the first period, which practically disappeared during the last period (Fig. 5.2d). This bloom was dominated by red-tide dinoflagellates, the toxic *Karenia mikimotoi*, and large *Ceratium* spp. (Chapter 8; Dahl and Johannessen, 1998). The onset of the spring bloom occurred earlier in the season (about 21 d) during the most recent period.

5.3.3 Copepods

The copepod time series are relatively short (Fig. 5.3). However, the abundance data suggest significantly decreasing trends in

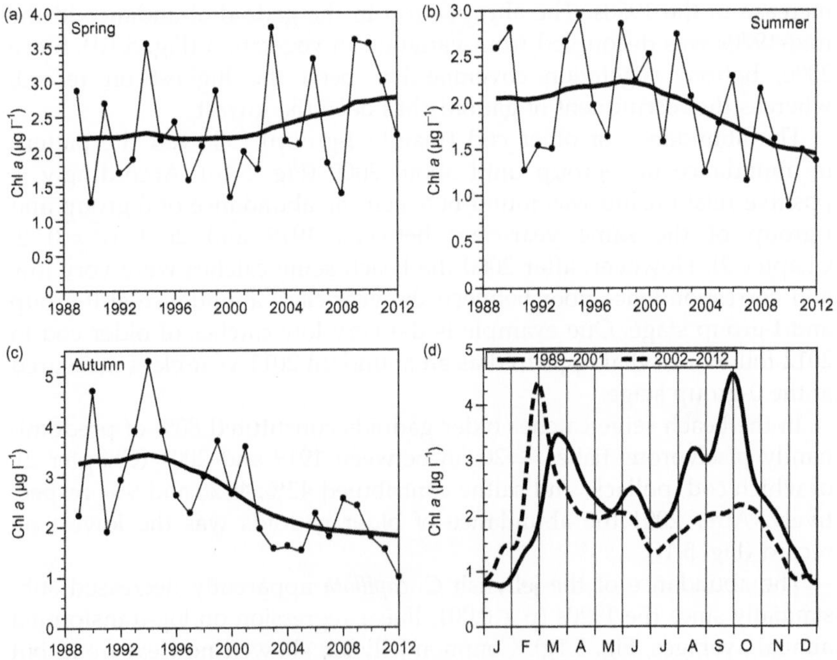

FIGURE 5.2 Mean Chl *a* in (a) spring; (b) summer; (c) autumn 1989–2012; (d) seasonal pattern in Chl *a* for the periods 1989–2001 and 2002–2012 (see Chapter 5 for more details). Trends in panels a, b, and c correspond to seven-year moving average estimated twice.

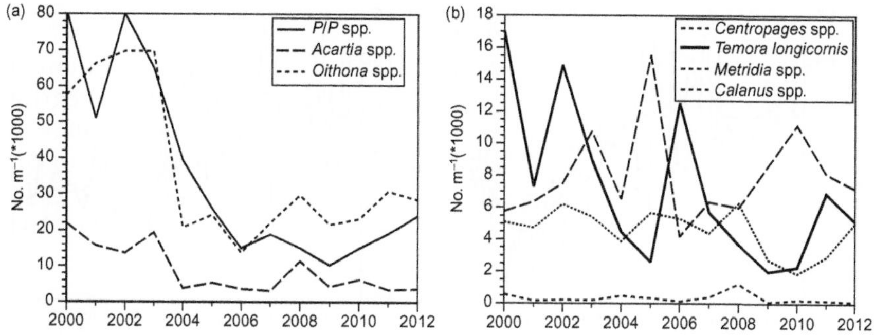

FIGURE 5.3 Annual means of copepods 2000–2012, (a) abundant (P/P–*Paracalans/Pseudocalanus*), (b) less abundant.

Paracalanus/Pseudocalanus ($p < 0.001$, linear regression), *Oithona* spp. ($p = 0.006$), *Acartia* spp. ($p = 0.013$), and *T. longicornis* ($p = 0.021$). For all these species the abundance seemed to shift from high to low around 2004. Reduction in abundance from the period 2000–2003 to 2004–2012 was 72% for *Acartia* spp., 71% for *Paracalanus/Pseudocalanus*, 64% for *Oithona* spp., and 58% for *Centropages* spp. The category *Calanus* spp. consists of *C. finmarchicus* and *C. helgolandicus*, two species that in wider long-term monitoring studies in the North Sea and Skagerrak showed decreasing and increasing trends, respectively (Reid et al., 2003).

During the Flødevigen time series (2000–2012), *Paracalanus/Pseudocalanus* was most abundant in July and August (Johannessen et al., 2012), coinciding with what appears to be the critical period for survival of 0-group cod in these waters (Chapters 3 and 4). The other species for which the abundances have been significantly reduced, *Oithona* spp., *Acartia* spp., and *T. longicornis*, also have peak abundances in July. *Calanus* spp. has peak abundance in April/May, *Metridia* spp. in May, and *Centropages* spp. in August-September (unpublished data IMR).

5.3.4 Oxygen

At intermediate depths (30 m) there was no trend in oxygen concentrations until the late 1960s (mean of all stations; Fig. 5.4a). In the following period, however, a linearly decreasing trend was observed, with the lowest in 2002. All individual stations showed negative trends between 1960 and 1993, and the majority were statistically significant (Johannessen and Dahl, 1996a). In 2003, oxygen increased abruptly ($p < 0.001$, STARS) to a higher level and subsequently fluctuated without any trend around this higher level.

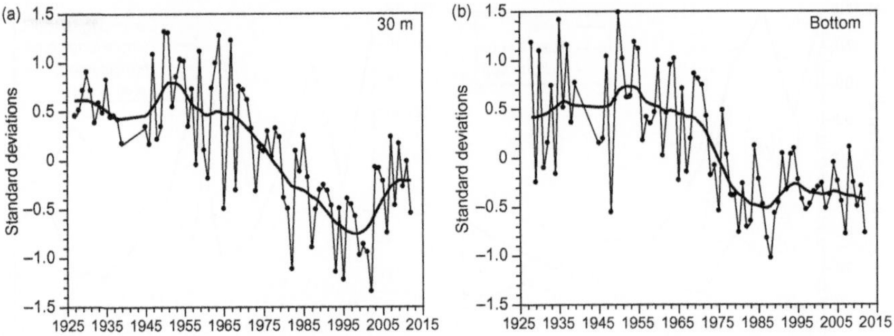

FIGURE 5.4 Historical trends in oxygen at (a) 30 m depth and (b) bottom waters (variable depth) at approximately 30 fixed stations along the Norwegian Skagerrak coast 1927–2012. Smoothed curves correspond to seven-year moving average estimated twice after standardization per station and depth.

In the mid-1970s there was an abrupt drop in bottom water oxygen ($p < 0.001$, STARS), with no trend before and after this shift (Fig. 5.4b).

5.3.5 Temperature

The annual mean surface temperature in Flødevigen bay (Fig. 5.5a) increased abruptly in the late 1980s ($p < 0.001$, STARS), from 8.9°C before 1988 to 10.0°C during the period 1988–2012. Biologically, the highest and lowest temperatures may have more impact than the average temperature. In winter there was an abrupt increase in temperature in the late 1980s (Fig. 5.5b), whereas the summer temperature increased gradually from the late 1980s (Fig. 5.5c). Both winter and summer temperatures decreased slightly in recent years.

5.4 DISCUSSION

5.4.1 Concurrent Shifts in the Plankton Community and Gadoid Recruitment Failures

The time series of 0-group gadoid abundance shows four major events (Fig. 5.1). The two first events, in the 1930s and in the 1950s, concurred with decreasing and increasing bottom flora coverage (Chapter 2). Bottom flora and particularly eelgrass meadows have long been recognized as important habitat for juvenile gadoids and other fishes (Blegvad, 1917). The results from the Skagerrak coast underline the significance of suitable habitat for recruitment of gadoids in coastal waters.

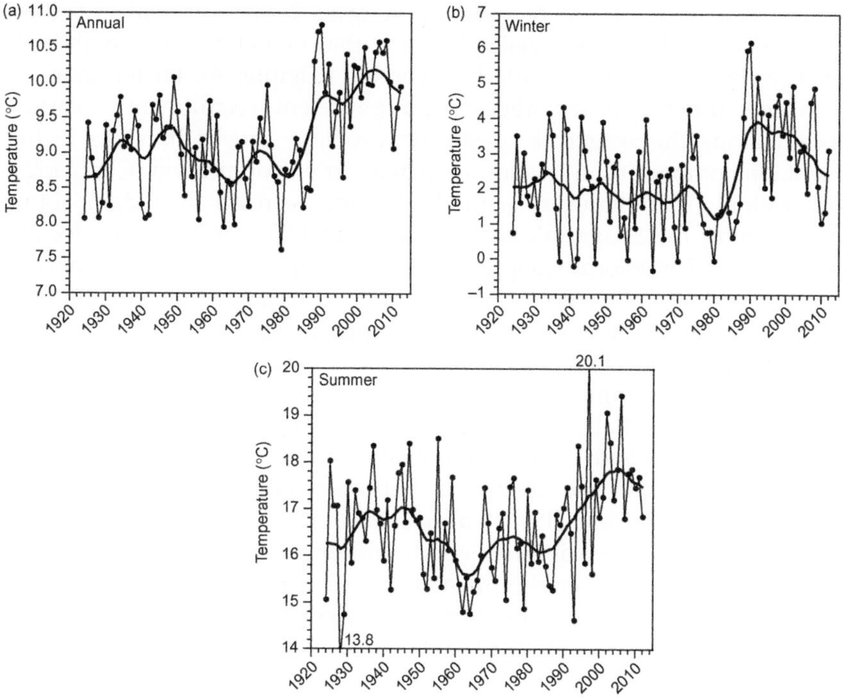

FIGURE 5.5 Mean sea surface temperature 1924–2012, (a) annual; (b) winter; (c) summer. Smoothed curves correspond to seven-year moving average estimated twice.

The two latter events, however, in the mid-1970s and in the early 2000s, were decoupled from changes in bottom vegetation. Bottom vegetation coverage has increased in recent years to the highest level on record, possibly because of improved light conditions resulting from reduced phytoplankton biomass (Fig. 5.2). This may allow eelgrass to grow at greater depths. Normally, the high bottom vegetation coverage should have been coupled with high abundance of young-of-the-year gadoids. However, the abundances of the gadoids after 2002 are the lowest on record, clearly suggesting that there are other mechanisms that underlie the extraordinary poor recruitment in the gadoids.

Comprehensive testing in the field suggested that the year-class strength in 0-group cod in these waters is determined after settlement, and it was concluded that predation was an unlikely cause of interannual variability in mortality rates (Chapters 3 and 4). The abundance of older gadoids (≥20 cm), that is, historically dominant members of the Skagerrak coastal fish community (Chapter 2), were the lowest on record after 2000 (Fig. 5.1e). Hence, intra- and interspecific predation is unlikely to be the cause of low abundances in the 0-group gadoids.

The field tests of cod recruitment suggested that survival in settled cod was limited by food availability, and that planktonic prey with high energy content were particularly favorable (Chapter 4). Furthermore, it was concluded that the abrupt and persistent recruitment collapses locally along the Skagerrak coast were probably caused by shifts in the plankton community in relation to increasing eutrophication (Chapter 2). Interestingly, simultaneous with the abrupt decrease in gadoid abundance along the Skagerrak coast in the mid-1970s, which were locally severe (e.g., Torvefjord, Chapter 2), abrupt drops in the oxygen concentrations of bottom waters occurred, with no trend before or after this event (Fig. 5.4b). A majority of the sampling stations are situated in fjords with sills in which the frequency of renewal of bottom water may vary from several times per year to several years between such events. As this may potentially obscure the rate of change in bottom water oxygen, Johannessen and Dahl (1996b) analyzed exposed stations without a sill and found that the change in bottom water oxygen was abrupt. Johannessen and Dahl (1996a) also analyzed a number of time series (temperature, salinity, density, wind, and precipitation) and found no simultaneous physical environmental change with that of oxygen in bottom waters or at intermediate depths. They concluded that reduced oxygen concentration was primarily related to increasing nutrient loads of the coastal water.

Wassmann (1991) reported that in inshore pelagic ecosystems sedimentation of phytoplankton and phytodetritus is extensive during periods of production when and if grazing rates by herbivores are low. In contrast, during heavy grazing by copepods, sedimentation of particulate material is much less, as copepod fecal pellets disintegrate and are recycled in the water column (Martens and Krause, 1990; Turner, 2002). Therefore, the abrupt drop in bottom water oxygen in the mid-1970s suggests that there was an abrupt change in the plankton community (Johannessen and Dahl, 1996a, 1996b) that led to a substantial increase in the sedimentation of organic matter (about 50% according to Aure et al., 1996). Hence, independent observations (fish and oxygen) point toward abrupt changes in the plankton community along the Norwegian Skagerrak coast in the mid-1970s. It should be noted though, that low oxygen was probably not the cause of recruitment failure in the gadoids but merely as signal of changes in the plankton community.

The decrease in recruitment of the gadoids in the early 2000s did not coincide with a decline in bottom water oxygen, but an abrupt increase in oxygen at intermediate depths (30 m). The oxygen level at intermediate depths may fluctuate substantially seasonally and is generally at a minimum in the latter half of September (Dahl and Danielssen, 1992). The linear decrease in oxygen up until 2002 probably reflected increased heterotrophic activity relative to primary productivity (Johannessen and Dahl, 1996a). Accordingly, the shift in 2003 suggests a reversal of the importance of these processes and is therefore indicative of abrupt changes in the plankton community. There was also a concurrent abrupt drop in autumn Chl *a* as a result of the intensive bloom of red tide dinoflagellates practically disappearing.

Lindahl and Hernroth (1983) reported that the autumn bloom of *Ceratium* spp. and *K. mikimotoi* started to appear around the mid-1970s. The decrease in bottom-water oxygen might have resulted from sedimentation of the relative intensive autumn bloom in following years. However, the substantial reduction in the autumn bloom after 2002 without changes in bottom-water oxygen concentration does not support this explanation.

The copepod time series is relatively short, and temporal trends, while rather clear, must be interpreted with some caution. There were abrupt decreases in the abundance of several species in the early 2000s. Among these are *Oithona* spp., which is a small (<1 mm) and primarily carnivorous species (Castellani et al., 2005; Nielsen and Sabatini, 1996). Carnivorous copepods have lower energy content than omnivorous and herbivorous copepods (Sargent and Falk-Petersen, 1988), and small copepods (<1 mm) in the diet of 0-group cod were associated with poor recruitment (Chapter 4). Hence, *Oithona* spp. is probably not important prey for the 0-group gadoids. The other copepods showing decreasing abundances, *Acartia* spp., *Temora longicornis* and *Paracalanus/Pseudocalanus* spp., are larger (1–2 mm) and omnivorous (Fileman et al., 2007; Peters et al., 2013). Their peak abundances are in July, concurrent with the critical period for survival in settled cod. Interestingly, *Paracalanus/Pseudocalanus*, which store lipids in oil sacs (Wlave and Larsson, 1999), have been reported to be important prey for early life stages of cod (Heath and Lough, 2007; Robert et al., 2011) and have also been associated with strong year-classes of cod in the North Sea (Rothschild, 1998).

Jellyfish and ctenophores are important consumers of zooplankton (Purcell, 2005). Although not quantified satisfactorily, there are no indications that ctenophores were ever abundant in the beach seine catches. *C. capillata* decreased substantially since the 1920s and the abundance was generally low after 2000. The abundance of common jellyfish fluctuated without trend, but the level after 2000 is the lowest on record. Hence, there is no evidence to suggest that jellyfish have contributed to reduced abundances in copepods in recent years.

The beach seine catches of ≥ I-group cod were extraordinary low after 2000, and also when seen in relation to the abundance at the 0-group stage. The reduction was particularly pronounced for the relatively strong 2011 year-class. One reason could have been cod migrating to deeper water due to higher temperatures in summer (Fig. 5.5c) and autumn. However, the summer temperature in 2012 was close to the average for the period 1924–1987 (16.8°C versus 16.4°C, 1 m depth), and so was the temperature during the beach seine survey in the latter half of September (14.2°C versus 14.0°C) that year. Hence, temperature per se is unlikely to be the reason for the 2011 year-class not showing up in the beach seine catches in 2012. In general, the temperature in the latter half of September (range 12.0°C to 16.8°C) after 2000 was within the optimal temperature range for cod in these waters. Therefore, the low catches of older cod after 2000 either reflect very low survival through the first winter or a behavioral response to some other factors, such as food availability in shallow water. Unfortunately, the present level of evidence cannot determine what is the mechanism behind the exceptionally low abundance of older cod. It worth noting though, that the decline in older cod was abrupt (Fig. 5.1d) and therefore yet another signal of abrupt ecosystem changes in the early 2000s.

The gadoid recruitment failure in the early 2000s may extend well beyond the Norway Skagerrak coast. In the North Sea, several of the most abundant species have had extraordinarily poor recruitment after 2000 (Alvarez-Fernandez et al., 2012; ICES, 2012; Payne et al., 2009), notably cod, herring (*Clupea harengus*), whiting, haddock (*Melanogrammus aeglefinus*), Norway pout (*Trisopterus esmarkii*), and lesser sandeel (*Ammodytes marinus*).

5.4.2 Ecosystem Bifurcations

There were recruitment collapses in the gadoids in the Inner Oslofjord around 1930 (Figs. 2.1, 2.9), in the Grenlandfjords and Holmestrandfjord in the mid-1960 (Figs. 2.7, 2.8), and a less severe recruitment failure generally along the Skagerrak coast in the mid-1970s (Fig. 5.1). Locally, the latter were as severe as in the three fjords described above. In addition, there were temporary reductions in gadoid abundances in relation to reduced coverage of bottom vegetation following the eelgrass disease in the early 1930s (Fig. 5.1d). The areas with persistent recruitment collapses showed common temporal patterns, and all events occurred during increasing eutrophication (Chapter 2). Comprehensive testing of the underlying mechanism provided evidence to suggest that the recruitment collapses were caused by abrupt shifts in the plankton community (Chapters 3 and 4). Likewise, coincident with the abrupt reductions in gadoid recruitment in the mid-1970s, the decrease in bottom-water oxygen concentration suggested abrupt changes in the plankton community as a result of increased nutrient loads (Johannessen and Dahl, 1996a, 1996b).

Nutrient concentration itself is not problematic for marine organisms. On the other hand, increased nutrient concentrations as well as altered nutrient composition will affect competition in phytoplankton (Anderson et al., 2002; Smayda, 1990). Eutrophication is generally a gradual process (Chapter 2). Hence, altered competition in the plankton community from gradual increasing nutrient loads is not reflected in the plankton community in the form of a gradual dose-response relationship but appears to reduce the resilience in the plankton community and thereby renders the community vulnerable to shift to alternative stable states. It is hypothesized that these shifts may be propagated to higher trophic levels by causing recruitment failure in fish. The new community structure after such shifts appears to be highly resilient, as there have been no signs of return to previous states after the shifts that took place 45 to 85 years ago. In the Grenlandfjords, the nutrient loads have been reduced well below the level at which the shift occurred (Johannessen and Dahl, 1996a).

As these shifts took place during gradually increasing eutrophication, several trophic levels were affected, and the shifts have been persistent. These shifts can be classified as bifurcations. In conclusion, the results suggest that marine plankton communities may shift abruptly from one stable state to another as a result of gradual environmental changes that are not detrimental to the organisms but affect competition in plankton.

The shift in the ecosystem in the early 2000s occurred during decreasing nutrient loads, as nitrate in the coastal current was reduced by approximately 30% from peak concentrations in 1991–1995 to 2001–2006 (Aure et al., 2010; Lindahl et al., 2009). During the same period, temperature increased. In winter, however, the temperature increased abruptly in the late 1980s, without causing an immediate response in the ecosystem that could be detected in the recruitment of gadoids or in oxygen concentrations. The summer temperature, on the other hand, increased gradually from the late 1980s. Despite occurring in relation to different environmental changes (eutrophication versus temperature), there are similarities between the ecosystem shifts as evidenced by shifts in the plankton community, which was propagated to higher trophic levels by causing recruitment failure in predator fishes, the gadoids. However, there are also a number of differences between the shifts.

In the early 2000s, the shift in oxygen was observed at intermediate depths with increasing oxygen concentrations, in contrast to decreasing oxygen concentrations in bottom waters in the mid-1970s. The intensive autumn blooms of red tide species started to appear in the mid-1970s (Lindahl and Hernroth, 1983), concomitant with the decrease in gadoid recruitment, whereas these blooms practically ceased in the early 2000s. The reduction in older gadoids ($\geq$ I-group) was less than that at the

0-group stage in the eutrophication-induced recruitment collapses (greater than 90% in 0-group versus approximately 70% in older gadoids in the Grenlandfjords and Holmestrandfjord; Chapter 3), whereas the abundance of particularly older cod was reduced much more than that of 0-group in the early 2000s (Fig. 5.1e).

An average temperature increase of 1.5°C in summer and winter is probably not problematic for the vast majority of plankton. However, like nutrients, temperature affects competition in plankton (Eppley, 1972; Goldman and Ryther, 1976; Stelzer 1998) and may thereby reduce the resilience and render the plankton community vulnerable to shifts. As the nutrient levels were still elevated when compared with pristine conditions (Aure et al., 2010; Lindahl et al., 2009), the shift in the early 2000s may have been caused by the aggregated effects of environmental changes that have altered competition in the plankton community. The response in the plankton community in the early 2000s differed from the earlier shifts though, which suggest that this shift was not a typically eutrophication response and therefore points toward temperature as an important factor.

Connell and Sausa (1983, p. 808) suggested a criterion to judge whether real ecosystems are stable: "the fate of all adults must...be followed for at least one complete overturn." The concurrent shifts in the plankton community and recruitment failure in the gadoids in the early 2000s occurred approximately 12 years ago, which according to this criterion can be considered as a persistent shift, as gadoids older than 10 years are extremely rare in these waters. Hence, the shift in the early 2000s appears to be a bifurcation. In agreement with this, Frigstad et al. (2013), who analyzed various time series from the Norwegian Skagerrak coast, found support for a regime shift around 2000 in these waters.

5.4.3 Regime Shifts

The literature on regime shifts has been growing steadily over the past three decades. However, the concept has become vague because of lack of consistent definitions of regimes and of regime shifts (deYoung et al., 2004; Overland et al., 2008). For example, Jarre et al. (2006) listed 13 different definitions of regime shifts from the marine literature.

Several of the regime shifts reported from large marine ecosystems have been liked to shifts in the physical environment, such as in the North Pacific (Hare and Mantua, 2000), in eastern boundary current systems (Chavez et al., 2003), and in the North Sea and the central Baltic Sea (Alheit et al., 2005). In a review of the North Pacific regime shifts, Overland et al. (2008, p. 100) state: "There is no convincing evidence that these climate shifts in the North Pacific are between multiple stable states. Instead, they appear to be more consistent with a long memory process with considerable autocorrelation at multiyear time scales, which can show persistent major deviations from a single century scale mean." In general, because of the difficulty of testing the underlying mechanism of shifts in large, open marine ecosystems (Collie et al., 2004; deYoung et al., 2004; Scheffer and van Nes, 2004), one can usually only come up with hypothesis of causal relationships.

One example from the Scotian Shelf on the east coast of Canada has been put forward as a particularly clear case of the ecosystem-wide impact of overfishing and climatic change (Scheffer et al., 2005; deYoung et al., 2008). Frank et al. (2005) found negative correlations between consecutive trophic levels and concluded that overfishing of benthic fishes (mainly cod) cascaded down the food web to nutrient concentrations (reduced biomass of benthic fish, increased biomass of benthic invertebrates and small pelagic fish, reduced biomass of zooplankton, increased biomass of phytoplankton, and reduced concentrations of nutrients). On the other hand, Choi et al. (2004) reported that the condition factor in benthic fishes decreased in parallel with these fishes being overfished. According to Frank et al. (2005, p. 1621), "The abundance of small pelagic fishes and benthic macroinvertebrates ...., once among the primary prey of the benthic fish community ..., increased markedly following the benthic fish collapse." Obviously, collapse in the benthic fishes coupled with reduced condition factor in these fishes and increased abundances in their primary prey are conflicting results.

Therefore, it remains unresolved whether regime shifts in large marine ecosystems, which are subjected to the combined impact of environmental change and substantial fishing pressure, merely reflect simple dose-response relationships from the environmental changes or whether high fishing pressures reduces the resilience and render the system vulnerable to shifts from environmental perturbations. To resolve these questions, there is a need for mechanistic understanding of how organisms in marine ecosystems interact and how important ecological processes are affected by environmental variability, environmental changes (e.g., climate), and human exploitation of marine resources.

Due to the general lack of tested mechanistic explanations, it cannot be determined whether the large-scale marine regime shifts were similar to the bifurcations observed along the Norwegian Skagerrak coast (i.e., shifts in the plankton community that were propagated to higher trophic levels by causing recruitment failure in fish). The only exception could be the shift in the plankton community around the turn of the century in the North Sea, which was associated with recruitment failure in a number of fishes (Alvarez-Fernandez et al., 2012; ICES, 2012).

As suggested, this shift could be a similar ecosystem response as the one observed along the Norwegian Skagerrak in relation to increasing temperature. Alvarez-Fernandez et al. (2012) suggested that climatic driven changes in the balance between nutrients could have been the cause of regime shift in the North Sea because the ratio between total soluble nitrogen (N) and phosphorous (P) dropped to below 20. However, it is questionable whether this drop could have severe ecological implications, as the natural N:P ratio in sea water is 15 (Redfield, 1958).

Prediction of ecological regime shifts is notoriously difficult (Biggs et al., 2009), but recent modeling studies indicate that changes in time series could potentially serve as early warning signals of imminent shifts (Guttal and Jayaprakash, 2008; Scheffer et al., 2009). These changes involve increased variance, skewness, autocorrelation, and return time after perturbations. However, as short-term (e.g., Lindahl and Perissinotto, 1987; Zagami et al., 1996) as well as long-term (e.g., Lindahl and Hernroth, 1988; Mackas et al., 2001) variability in the plankton community is very high, it seems unlikely that the proposed warning signals can be used to evaluate if shifts in plankton communities are impending. In addition, a modeling study by Boerlijst et al. (2013) suggests that regime shifts may occur without early warnings.

5.5 CONCLUSION

Because of inconsistent use of the regime shift concept, the term "ecosystem bifurcation" has been used here to be more precise about the kind of regime shifts considered that is, abrupt and persistent shifts that affect several trophic levels and that are a result of gradual environmental changes. The question of shifts between contrasting states is controversial, and abrupt ecosystem shifts are counterintuitive to many (Scheffer and Carpenter, 2003), in particular the idea of bifurcations. One reason for this skepticism is probably the lack of repeated documented incidents from real ecosystems to confirm that observed shifts are results of gradual environmental changes. However, the unique beach seine time series from the Norwegian Skagerrak coast have shown repeated incidents of abrupt and persistent recruitment failures in gadoids (Fig. 5.1, Fig. 2.6, Fig. 2.7, Fig. 2.8, and Fig. 2.9). Comprehensive testing in the field and parallel time series on oxygen and plankton have provided evidence to suggest that these recruitment failures were linked to abrupt changes in the plankton community, both in relation to increasing nutrient loads and increasing temperature. It has been suggested that the changing environmental conditions alter the competition in plankton and thereby reduce the resilience of the system, which becomes vulnerable to shift to an alternative stable state from biological or environmental perturbations that the system could withstand under optimal environmental conditions (Fig. 1.1). According to this interpretation of bifurcations, the altered plankton community will consist of organisms for which the environmental conditions are closest to optimal. The altered community structures after such shifts appear to be highly resilient, as there have been no signs of recovery, even when the environmental conditions have returned to levels well below that at which the shifts occurred, as with hysteresis (e.g., Andersen et al., 2009).

Another probable reason for the skepticism toward ecosystem bifurcations is the difficulty of identifying mechanistic explanations for resilience in systems without obvious keystone species or keystone ecological processes. In Chapter 6, a mechanistic explanation for resilience in the plankton community is proposed.

References

Alheit, J., Möllmann, C., Dutz, J., Kornilovs, G., Loewe, P., Mohrholz, V., et al., 2005. Synchronous ecological regime shifts in the central Baltic and the North Sea in the late 1980s. ICES J. Mar. Sci. 62, 1205–1215.

Alvarez-Fernandez, S., Lindeboom, H., Meesters, E., 2012. Temporal changes in plankton of the North Sea: community shifts and environmental drivers. Mar. Ecol. Prog. Ser. 462, 21–38.

Andersen, T., Carstensen, J., Hernández-García, E., Duarte, C.M., 2009. Ecological thresholds and regime shifts: approaches to identification. Trends Ecol. Evol. 24, 49–57.

Anderson, D.M., Glibert, P.M., Burkholder, J.M., 2002. Harmful algal blooms and eutrophication: nutrient sources, composition, and consequences. Estuaries 25, 704–726.

Aure, J., Danielssen, D.S., Sætre, R., 1996. Assessment of eutrophication in Skagerrak coastal waters using oxygen consumption in fjordic basins. ICES J. Mar. Sci. 53, 589–595.

Aure, J., Danielssen, D., Magnusson, J., 2010. Langtransporterte tilførsler av næringssalter til Ytre Oslofjord 1996–2006. Fisken og havet 4/2010, 1–24 (in Norwegian).

Biggs, R., Carpenter, S.R., Brock, W.A., 2009. Turning back from the brink: detecting an impending regime shift in time to avert it. Proc. Natl. Acad. Sci. USA. 106, 826–831.

Blegvad, H., 1917. Om fiskenes føde i de danske farvande innden for Skagen. Beretning til Landbrugsministeriet 24, 17–72 (in Danish).

Boerlijst, M.C., Oudman, T., de Roos, A.M., 2013. Catastrophic collapse can occur without early warning: examples of silent catastrophes in structured ecological models. PLoS One 8, e62033.

Castellani, C., Irigoien, X., Harris, R.P., Lampitt, R.P., 2005. Feeding and egg production of *Oithona similis* in the North Atlantic. Mar. Ecol. Prog. Ser. 288, 173–182.

Charpy-Roubaud, C., Sournia, A., 1990. The comparative estimation of phytoplanktonic, microphytobenthic and macrophytobenthic primary production in the oceans. Mar. Microb. Food Webs 4, 31–57.

Chavez, F.P., Ryan, J., Lluch-Cota, S.E., Ñiquen, M., 2003. From anchovies to sardines and back: multidecadal change in the Pacific Ocean. Science 299, 217–221.

Choi, J.S., Frank, K.T., Leggett, W.C., Drinkwater, K., 2004. Transition to an alternate state in a continental shelf ecosystem. Can. J. Fish. Aquat. Sci. 61, 505–510.

Collie, J.S., Richardson, K., Steele, J.H., 2004. Regime shifts: can ecological theory illuminate the mechanisms? Prog. Oceanogr. 60, 281–302.
Connell, H., Sausa, W.P., 1983. On the evidence needed to judge eocological stability or persistence. Am. Nat. 121, 789–824.
Dahl, E., Danielssen, D.S., 1992. Long-term observations of oxygen in the Skagerrak. ICES Mar. Sci. Symp. 195, 455–461.
Dahl, E., Johannessen, T., 1998. Temporal and spatial variability of phytoplankton and Chl a—lessons from the south coast of Norway and the Skagerrak. ICES J. Mar. Sci. 55, 680–687.
deYoung, B., Harris, R., Alheit, J., Beaugrand, G., Mantua, N., Shannon, L., 2004. Detecting regime shifts in the ocean: data considerations. Prog. Oceanogr. 60, 143–164.
deYoung, B., Barange, M., Beaugrand, B., Harris, R., Perry, R.I., Scheffer, M., et al., 2008. Regime shifts in marine ecosystems: detection, prediction and management. Trends. Ecol. Evol. 23, 402–409.
Eppley, R.W., 1972. Temperature and phytoplankton growth in the sea. Fish. Bull. 1063–1085.
Fileman, E., Smith, T., Harris, R., 2007. Grazing by *Calanus helgolandicus* and *Para-Pseudocalanus* spp. on phytoplankton and protozooplankton during the spring bloom in the Celtic Sea. J. Exp. Mar. Biol. Ecol. 348, 70–84.
Frank, K.T., Petrie, B., Choi, J.S., Leggett, W.C., 2005. Trophic cascades in a formerly cod-dominated ecosystem. Science 308, 1621–1623.
Frigstad, H., Andersen, T., Hessen, D.O., Jeansson, E., Skogen, M., Naustvoll, L.-J., et al., 2013. Long-term trends in carbon, nutrients and stoichiometry in Norwegian coastal waters: evidence of a regime shift. Prog. Oceanogr. 111, 113–124.
Goldman, J.C., Ryther, J.H., 1976. Temperature-influenced species competition in mass cultures of marine phytoplankton. Biotech. Bioengr. 18, 1125–1144.
Gunderson, L.H., 2000. Ecological resilience—in theory and application. Annu. Rev. Ecol. Syst. 31, 425–439.
Guttal, V., Jayaprakash, C., 2008. Changing skewness: an early warning signal of regime shifts in ecosystems. Ecol. Lett. 11, 450–460.
Hare, S.R., Mantua, N.J., 2000. Empirical evidence for North Pacific regime shifts in 1977 and 1989. Prog. Oceanogr. 47, 103–145.
Heath, M.R., Lough, R.G., 2007. A synthesis of large-scale patterns in the planktonic prey of larval and juvenile cod (*Gadus morhua*). Fish. Oceanogr. 16, 169–185.
Holling, C.S., 1973. Resilience and stability of ecological systems. Annu. Rev. Ecol. Syst. 4, 385–398.
Hutchinson, G., 1961. The paradox of the plankton. Am. Nat. 95, 137–145.
ICES, 2012. Report of the ICES Advisory Committee 2012. ICES Advice, 2012. Book 6.
Jarre, A., Moloney, C.L., Shannon, L.J., Fréon, P., van der Lingen, C.D., Verheye, H.M., et al., 2006. Developing a basis for detecting and predicting long-term ecosystem changes. In: Shannon, V., Hempel, G., Malanotte-Rizzoli, P., Moloney, C., Woods, J. (Eds.), Benguela: Predicting a large Marine Ecosystem. Large Marine Ecosystem Series 14. Elsevier, Amsterdam, pp. 239–272.
Johannessen, T., Dahl, E., 1996a. Declines in oxygen concentrations along the Norwegian Skagerrak coast 1927–1993: a signal of ecosystem changes due to eutrophication? Limnol. Oceanogr. 41, 766–778.
Johannessen, T., Dahl, E., 1996b. Historical changes in oxygen concentrations along the Norwegian Skagerrak coast: a reply to the comment by Gray and Abdullah. Limnol. Oceanogr. 41, 1847–1852.
Johannessen, T., Dahl, E., Falkenhaug, T., Naustvoll, L.J., 2012. Concurrent recruitment failure in gadoids and changes in the plankton community along the Norwegian Skagerrak coast after 2002. ICES J. Mar. Sci. 69, 795–801.

Leis, J.M., 2007. Behaviour as input for modelling dispersal of fish larvae: behaviour, biogeography, hydrodynamics, ontogeny, physiology and phylogeny meet hydrography. Mar. Ecol. Prog. Ser. 347, 185–193.

Lewontin, R.C., 1969. The meaning of stability. Brookhaven Symp. Biol. 22, 13–24.

Lindahl, O., Hernroth, L., 1983. Phyto-zooplankton community in coastal waters of western Sweden—an ecosystem off balance? Mar. Ecol. Prog. Ser. 10, 119–126.

Lindahl, O., Hernroth, L., 1988. Large-scale and long-term variations in the zooplankton community of the Gullmar Fjord, Sweden, in relation to advective processes. Mar. Ecol. Prog. Ser. 43, 161–171.

Lindahl, O., Perissinotto, R., 1987. Short-term variations in the zooplankton community related to water exchange processes in the Gullmar Fjord. Sweden. J. Plankton Res. 9, 1113–1132.

Lindahl, O., Andersson, L., Belgrano, A., 2009. Primary phytoplankton productivity in the Gullmar Fjord, Sweden. An evaluation of the 1985–2008 time series. The Swedish Environmental Protection Agency, Report 3606, 1–35.

Mackas, D.L., Thomson, R.E., Galbraith, M., 2001. Changes in the zooplankton community of the British Columbia continental margin, 1985–1999, and their covariation with oceanographic conditions. Can. J. Fish. Aquat. Sci. 58, 685–702.

Martens, P., Krause, M., 1990. The fate of faecal pellets in the North Sea. Helgol. Wiss. Meeresunters. 44, 9–19.

May, R.M., 1977. Thresholds and breakpoints in ecosystems with a multiplicity of stable states. Nature 269, 471–477.

Motoda, S., 1959. Devices of simple plankton apparatus. Mem. Fac. Fish. Hokkaido Univ. 7, 73–94.

Nielsen, T.G., Sabatini, M., 1996. Role of cyclopoid copepods *Oithona* spp. in North Sea plankton communities. Mar. Ecol. Prog. Ser. 139, 79–93.

Overland, J., Rodionov, S., Minobe, S., Bond, N., 2008. North Pacific regime shifts: definitions, issues and recent transitions. Prog. Oceanogr. 77, 92–102.

Paine, R.T., 1969. A note on trophic complexity and community stability. Am. Nat. 103, 91–93.

Payne, M.R., Hatfield, E.M., Dickey-Collas, M., Falkenhaug, T., Gallego, A., Gröger, J., et al., 2009. Recruitment in a changing environment: the 2000s North Sea herring recruitment failure. ICES J. Mar. Sci. 66, 272–277.

Pechenik, J.A., 1999. On the advantages and disadvantages of larval stages in benthic marine invertebrate life cycles. Mar. Ecol. Prog. Ser. 177, 269–297.

Peters, J., Dutz, J., Hagen, W., 2013. Trophodynamics and life-cycle strategies of the copepods *Temora longicornis* and *Acartia longiremis* in the Central Baltic Sea. J. Plankton Res. 35, 595–609.

Pimm, S.L., 1991. The Balance of Nature? University of Chicago Press, London.

Purcell, J.E., 2005. Climate effects on formation of jellyfish and ctenophore blooms: a review. J. Mar. Biol. Ass. U.K. 85, 461–476.

Redfield, A.C., 1958. The biological control of chemical factors in the environment. Am. Sci. 46, 205–221.

Reid, P.C., Edwards, M., Beaugrand, G., Skogen, M., Stevens, D., 2003. Periodic changes in the zooplankton of the North Sea during the twentieth century linked to oceanic inflow. Fish. Oceanogr. 12, 260–269.

Robert, D., Levesque, K., Gagné, J.A., Fortier, L., 2011. Change in prey selectivity during the larval life of Atlantic cod in the southern Gulf of St Lawrence. J. Plankton Res. 33, 195–200.

Rodionov, S., Overland, J., 2005. Application of a sequential regime shift detection method to the Bering Sea ecosystem. ICES J. Mar. Sci. 62, 328–332.

Rothschild, B.J., 1998. Year class strength of zooplankton in the North Sea and their relation to cod and herring abundance. J. Plankton Res. 20, 1721–1741.
Sargent, J.R., Falk-Petersen, S., 1988. The lipid biochemistry of calanoid copepods. Hydrobiologia 167/168, 101–114.
Scheffer, M., Carpenter, S., 2003. Catastrophic regime shifts in ecosystems: linking theory to observations. Trends Ecol. Evol. 18, 648–656.
Scheffer, M., van Nes, E.H., 2004. Mechanisms for marine regime shifts: can we use lakes as microcosms for cceans? Prog. Oceanogr. 60, 303–319.
Scheffer, M., Carpenter, S., Foley, J., Folke, C., Walker, B., 2001. Catastrophic shifts in ecosystems. Nature 413, 591–596.
Scheffer, M., Carpenter, S., de Young, B., 2005. Cascading effects of overfishing marine systems. Trends Ecol. Evol. 20, 579–581.
Scheffer, M., Bascompte, J., Brock, W.A., Brovkin, V., Carpenter, S.R., Dakos, V., et al., 2009. Early-warning signals for critical transitions. Nature 461, 53–59.
Short, F.T., Ibelings, B.W., den Hartog, C., 1988. Comparison of a current eelgrass disease to the wasting disease in the 1930s. Aquat. Bot. 30, 295–304.
Smayda, T., 1990. Novel and nuisance phytoplankton blooms in the sea: evidence for a global epidemic. In: Graneli, E., Sundstrom, B., Edler, L., Anderson, D.M. (Eds.), Toxic Marine Phytoplankton. Elsevier, New York, pp. 29–40.
Stelzer, C.-P., 1998. Population growth in Planktonic Rotifers. Does temperature shift the competitive advantage for different species? Hydrobiologia 387/388, 349–353.
Strickland, J.D.H., Parsons, T.R., 1968. A practical handbook of seawater analysis. Fish. Res. Bd. Can. Bull. 167, 1–317.
Thorson, G., 1950. Reproductive and larval ecology of marine bottom invertebrates. Biol. Rev. 25, 1–4.
Turner, J., 2002. Zooplankton fecal pellets, marine snow and sinking phytoplankton blooms. Aquat. Microb. Ecol. 27, 57–102.
UNESCO, 1968. Zooplankton Sampling. UNESCO, Paris, 174 p.
Wassmann, P., 1991. Dynamics of primary productivity and sedimentation in shallow fjords and polls of western Norway. Oceanogr. Mar. Biol. Annu. Rev. 29, 87–154.
Wlave, J., Larsson, U., 1999. Carbon, nitrogen and phosphorus stoichiometry of crustacean zooplankton in the Baltic Sea: implications for nutrient recycling. J. Plankton Res. 21, 2309–2321.
Zagami, G., Badalamenti, F., Guglielmo, L., Manganaro, A., 1996. Short-term variations of the zooplankton community near the Straits of Messina (North-Eastern Sicily): relationships with the hydrodynamic regime. Estuar. Coast. Shelf Sci. 42, 667–681.

CHAPTER

6

Predator-Prey Synergism in Plankton

6.1 INTRODUCTION

6.1.1 Predator-Prey Relationship

Here, the words predator and prey are used in the broadest sense of the words, including grazers as predators on primary producers. There is growing evidence from both aquatic and terrestrial ecosystems that the relationships between primary producers and herbivores are complex and include both the direct impact of grazing and the indirect impact of the recycling of nutrients (Elser and Urabe, 1999; McNaughton et al., 1997; Sterner, 1986). Nevertheless, it would seem justified to assert that it is still an important assumption in ecological theory that interactions between predators and prey are mainly negative (Loreau, 1995), implying that a high abundance of a predator reduces the abundance of its prey. From the perspective of systems ecology, this assumption is linked to the perception that nutritional requirements cause organisms to compete for resources and/or to feed on each other, leading to negative interactions between populations (competition, predation, parasitism), with symbiosis as a rather exotic case (Sommer, 1989). Also, from a reductionist perspective (the level of the individual), the predator-prey relationship is obviously negative because the predator either kills or damages its prey. A potential problem with both of these perspectives is that only the direct predator-prey relationship is considered. It is conceivable that even though a grazer has a negative impact on its preferred plants seen in comparison with specimens of the same species that are not being grazed, the preferred plants may gain competitive advantages over non-grazed species by being only mildly affected by the grazing

From an Antagonistic to a Synergistic Predator Prey Perspective.
DOI: http://dx.doi.org/10.1016/B978-0-12-417016-2.00006-3

and by the grazer weeding out non-grazed competitors. McNaughton (1979) calls this competitive fitness.

In limnic ecology the generally accepted perception of a negative predator-prey relationship is reflected in the trophic cascade hypothesis, among others (Carpenter et al., 1985). This hypothesis, which in principal is the same as the "world is green" hypothesis originally proposed for terrestrial systems by Hairston et al. (1960), states that in the absence of a higher trophic level, top predators compete for limited resources and thereby reduce the abundance of their prey below their carrying capacity, which in turn causes the third highest trophic level to compete for limited resources. This alternating control of trophic levels via competition and predation cascades down to the base of the food web.

In marine ecosystems, it is widely accepted that phytoplankton are controlled by grazers, whereas algal blooms are more occasional events (Legendre, 1990; Riegman et al., 1993; Thingstad and Sakshaug, 1990). Accordingly, much attention is paid to the influence grazing may have on the diversity and structure of phytoplankton communities (e.g., Thingstad, 1998) and under which conditions phytoplankton may escape grazing control and bloom (Irigoien et al., 2005; Mitra and Flynn, 2006; Smayda, 1997a). An obvious prerequisite for grazing control is that edible phytoplankton predominate. Results reported from various time series collected in the coastal waters of Skagerrak (Fig. 6.1) support the perception of grazing control and the predominance of edible algae during a substantial part of the productive season. However, rather than taking palatability in phytoplankton for granted and analyze the impact of grazing on the algal community from this perspective, the question about how evolution apparently has favored edible algae over low- and non-grazed species under high grazing pressure is addressed. A mechanism is proposed to account for this phenomenon, in which the cycling of nutrients is a key ecological process. Edible phytoplankton gain competitive advantages by "sacrificing" parts of their clonal populations in order to obtain resources for continuous growth. The same applies to bacteria and heterotrophic nanoflagellates (HNF), which reproduce asexually and are key players in the cycling of nutrients (Azam et al., 1983), and to phages (virus "predating" bacteria) as well. The proposed mechanism is a win-win situation for zooplankton and phytoplankton (and bacteria, HNF, and phages) because the grazers stimulate production of their preferred algal prey. As phytoplankton, bacteria, and HNF sacrifice only parts of their clonal populations, genetically identical cells continue to grow and multiply. Hence, the proposed mechanism is in agreement with the perception of the "selfish gene." The idea of a positive relationship between herbivores and primary producers is also discussed in relation to some terrestrial systems.

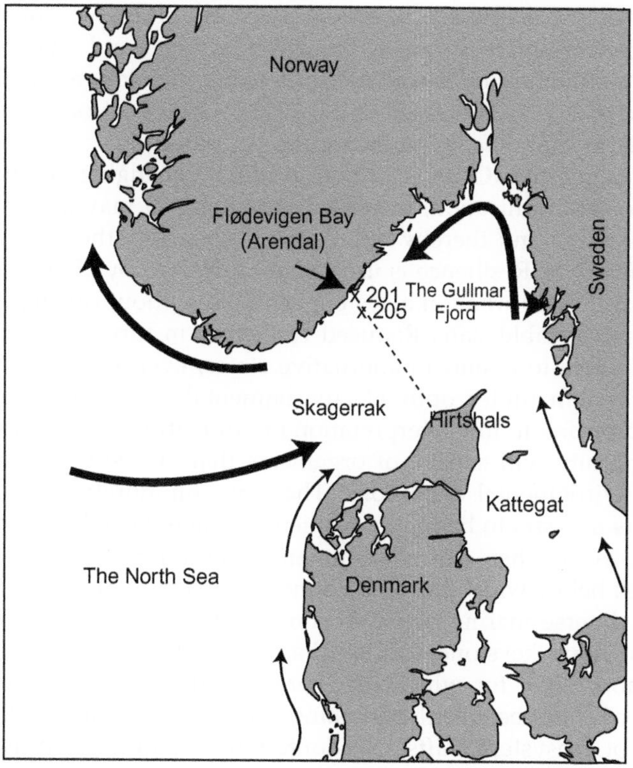

FIGURE 6.1 Skagerrak is a non-tidal area with a marked stratified water-column during most of the year. Arrows indicate the main surface circulation pattern. Chl *a* was measured in Flødevigen Bay, nutrients at stations 201 and 205 situated 1 and 5 nautical miles offshore, respectively, and primary productivity in the mouth of Gullmar fjord.

6.1.2 Ecosystem Bifurcations and Resilience

Another motivation for questioning the traditional perception of a negative predator-prey relationship stems from the search for a mechanism to explain repeated incidents of ecosystem bifurcations locally along the Norwegian Skagerrak coast. Bifurcations are abrupt and persistent regime shifts that affect several trophic levels and are caused by gradual environmental changes. The shifts became apparent through sudden and persistent recruitment failures in gadoid fishes (Chapter 2) and sudden decreases in bottom water oxygen, indicating structural changes in the plankton community in relation to gradually increasing anthropogenic eutrophication (Chapter 5; Johannessen and Dahl, 1996a, 1996b). Comprehensive testing in the field provided evidence suggesting that the recruitment collapses were caused by abrupt changes in the plankton community that deprived young-of-the year gadoids of adequate prey (Chapters 2, 3, and 4). Recently, there was another incident of concurrent recruitment failure in gadoids and changes in the plankton community on a regional scale along the Norwegian Skagerrak coast (Chapter 5; Johannessen et al., 2012). This event occurred during decreasing nutrient loads but increasing temperature.

It was suggested (Chapter 5) that gradually changing environmental conditions (eutrophication, increased temperature) may affect competition in plankton and thereby reduce the resilience in the plankton community (Fig. 1.1). Resilience is used here as defined by Holling (1973) as the maximum perturbation a system can take without causing a shift to an alternative stable state. Reduced resilience, in turn, renders the system vulnerable to a shift to alternative states, even from perturbations that the system under optimal environmental conditions could withstand. According to this interpretation of bifurcations, the altered plankton community will consist of organisms that are best adapted to the altered environmental conditions. The new community structure after such shifts appears to be highly resilient, as there have been no signs of recovery, even when the environmental conditions have returned to levels well below which the shifts occurred (Chapter 5).

In the diverse marine plankton community (Hutchinson, 1961) there are no obvious keystone species (Paine, 1969) or keystone processes (Gunderson, 2000) for which there are tolerance limits. Therefore, the question of shifts between contrasting states still remains controversial, and abrupt ecosystem shifts are counterintuitive to many (Scheffer and Carpenter, 2003), in particular the idea of bifurcations. There are probably two reasons for this skepticism. One reason is lack of repeated incidents to confirm that shifts result from gradual environmental changes. However, the repeated incidents of abrupt and persistent recruitment failures in gadoids in relation to gradually changing environmental conditions along the Norwegian Skagerrak coast, which could be linked to abrupt changes in the plankton community, do suggest that marine ecosystems are vulnerable to bifurcations (Chapter 2–5). The second reason for the skepticism is the problem to come up with mechanistic explanations for resilience in systems without apparent keystone species or keystone processes. In this chapter, I propose that the positive coexistence of organisms in the planktonic food web may provide a mechanistic explanation for resilience, and I argue that high resilience is incompatible with negative predator-prey relationships (antagonism).

6.2 METHODS

This study was carried out in Skagerrak (Fig. 6.1), which is a nontidal area with a vertically stratified water column. It is a transit area for water masses flowing from the Baltic, Kattegat, and the North Sea to waters farther north. Data were collected over the ten-year period 1990–1999. Nutrient concentrations (nitrate, phosphate, and silicate) were measured biweekly at station 201 (Fig. 6.1) at depths of 0, 5, 10, 20, 30, 50, and 75 m, and monthly at station 205 at the same depths as for 201, but at depths of 100, 125, 150, and 240 m as well. Analysis of nutrients was performed according to standard procedures (Hansen and Koroleff, 1999). Chl *a* and primary productivity measurements between 1990 and 1999 were modified from Johannessen et al. (2006). Chl *a* at 0–3 m depth was measured three times a week in Flødevigen Bay on the adjacent Skagerrak coast of Norway (Fig. 6.1). Primary productivity was measured in situ by the 14C-incorporation technique at ten depths (0, 1, 2, 3, 4, 6, 8, 10, 15, and 20 m) in the mouth area of the Gulmar Fjord on the Swedish west coast (Fig. 6.1). Dark-bottle incubations were performed at 0 and 20 m depth. The incubations were carried out over a period of four hours around noon (for more details see Lindahl, 1995; Johannessen et al., 2006).

6.3 RESULTS

Chl *a*, primary productivity and nutrient concentrations at different depths pooled for the 10-year period 1990–1999, are presented in Fig. 6.2. The algal biomass as measured in terms of Chl *a* is characterized by a spring bloom in March, relatively low biomass from April to July, followed by a long and intensive autumn bloom that peaks in September–October. Primary productivity increases and nutrients decrease in parallel with the onset of the spring bloom. Silicate is utilized mainly during the spring bloom (Fig. 6.2d), which supports the classical view that diatoms dominate vernal primary production. Nitrate is utilized deeper in the water column than either silicate or phosphorous are, which supports the general perception that nitrogen limits primary production in marine ecosystems (Ryther and Dunstan, 1971). After the spring bloom, primary productivity continues to increase and reaches a maximum in summer as a result of the gradual utilization of nutrients at greater depths (see nitrate, Fig. 6.2b). Nutrients from zooplankton exudation and excretion are reutilized by phytoplankton (Sterner, 1986). When the import of new nutrients has come to a halt around July, primary production is based on recycled nutrients. When primary productivity starts to decrease in late summer, nutrients are lost from the euphotic zone, as evidenced by increasing nitrate concentrations greater than 75 m. The decrease in primary productivity coincides with the start of the autumn bloom. Hence, there is a tendency toward a negative correlation between primary productivity

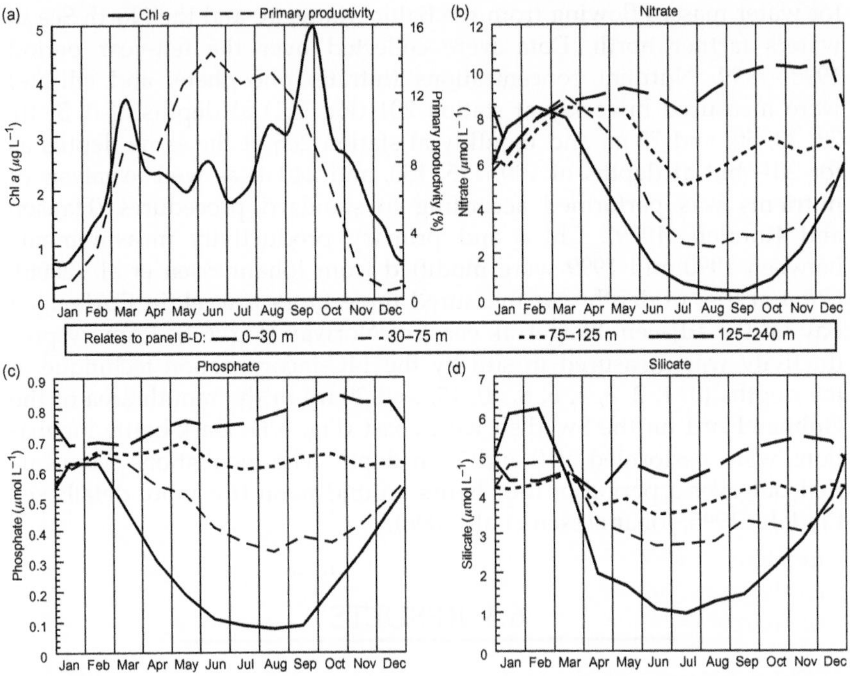

FIGURE 6.2 Seasonal pattern in (a) Chl *a* and primary productivity and nutrients at depth intervals, (b) nitrate, (c) phosphate, and (d) silicate for the period 1990–1999 in Skagerrak coastal waters (Fig. 6.1). Primary productivity is expressed in terms of monthly contributions (%) to annual primary productivity. Panel a is reproduced from Johannessen et al. (2006), with permission from Taylor & Francis.

and algal biomass during the productive season (r = −0.30; Pearson correlation coefficient between monthly means March–October).

At the onset of the autumn bloom, nitrate is depleted over the 0–30 m depth range, primary productivity decreases, and nutrients are lost from the euphotic zone as observed by increasing concentrations in deeper waters. Since the nutrient data represent averages over relatively wide ranges of depths, and the time of onset of the blooming varies considerably from year to year (Dahl and Johannessen, 1998), detailed information relevant to the individual autumn bloom might potentially be obscured. Therefore, in Fig. 6.3 nitrate concentrations in the upper 75 m of the water column measured just before the autumn blooms (average 2 days) are compared with concentrations measured during the subsequent sampling (average 15 days after the onset of the blooms). At 0–20 m depth, nitrate levels just before the outburst are the same as those measured on the date of the next sampling, whereas nitrate levels at 50 and 75 m are slightly lower prior to the blooms. The indication of

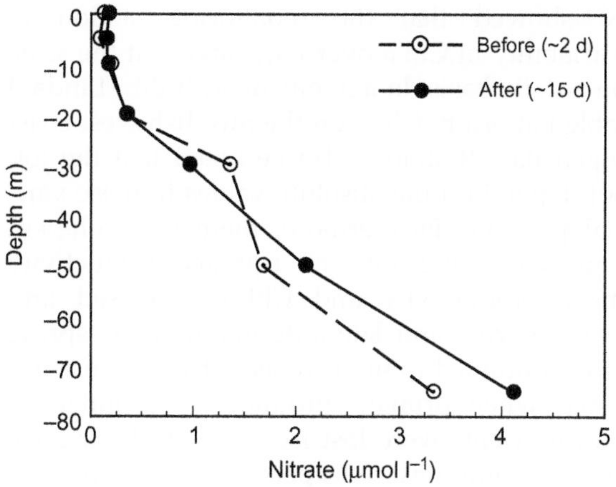

FIGURE 6.3 Average nitrate concentrations at 0–75 m depth at station 201 before (average 2 days) and after (average 15 days) the commencement of the autumn algal bloom for the period 1990–1999.

higher values before the bloom at 30 m depth is generated entirely by one outlier, which contributed more than the sum of the remaining measurements. With this outlier excluded, the nitrate level at 30 m depth is slightly lower prior to the blooms. This observation further emphasizes that inorganic nutrients are depleted in the surface layer and nutrients are lost from the euphotic zone at the onset of the autumn bloom. Furthermore, precipitation does not increase notably from July to August and September (Johannessen et al., 2006). Hence, there is no evidence of new nutrient supply triggering the autumn outburst. These results support neither the classical view that the autumn bloom is a result of new nutrients being mixed into the euphotic zone by, for example, storm events, nor the view that nutrients are being utilized below the euphotic zone by vertically migrating algae. Rather, the generally intensive autumn bloom in these waters is based on recycled nutrients.

6.4 DISCUSSION

6.4.1 Methodological Considerations

There are potentially two methodological problems inherent in this study: (1) Chl *a* and nutrient concentrations were measured on the south coast of Norway, whereas primary productivity was measured on the west coast of Sweden, and (2) the annual pattern in Chl *a* was described at the surface (0–3 m), whereas primary productivity was measured over the 0–20 m depth layer (Johannessen et al., 2006). The water masses in the sampling areas on the west coast of Sweden and the south coast of Norway are principally the same on account of the prominent coastal current from east to west (Sætre, 2007). Also, Dahl and Johannessen (1998) showed that the monitoring station in Flødevigen Bay reflects variability in Chl *a* over large areas (at the scale of the Skagerrak), even on a daily basis. In agreement with this, Lindahl (1995) reported a seasonable pattern in Chl *a* on the Swedish west coast similar to that in Flødevigen Bay. It should also be noted that the following reasoning does not depend on the absolute values in these variables but on the seasonal patterns. Furthermore, there is a coherent pattern between Chl *a*, nutrient utilization, and primary productivity (Fig. 6.2), that is, primary productivity and Chl *a* increased and decreased in parallel in early spring and late autumn, after the spring bloom primary productivity continued to increase as nutrients were utilized gradually deeper, and when primary productivity commenced decreasing in late summer, nutrients were lost from the euphotic zone as evidenced by increasing concentrations in deeper waters (Figs. 6.2 and 6.3). In addition, close correspondence ($r^2 = 0.99$) of average monthly (January–December) Chl *a* concentration at the surface (0–5 m) and in the upper 20 m depth layer at a station just outside Flødevigen Bay suggests that the pattern observed in surface Chl *a* is representative of a wider depth range (r^2 was estimated from monthly measurements over a 20-year period). Consequently, it can be concluded that despite geographical distance and measurements obtained from different depth layers, the comparison of patterns in Chl *a* and primary productivity is justified.

6.4.2 Grazing Control

To account for the apparent paradox of the negative correlation between phytoplankton biomass and primary productivity, the following conceptual model of primary productivity can be used (Johannessen et al., 2006):

$$P = \Delta A + S + G + E \tag{6.1}$$

where P is production of carbon, ΔA represents changes (positive and negative) in algal biomass (biomass of carbon per unit of time), S is sedimentation of algal carbon, G is grazing of algal carbon (including lysis of cells followed by bacterial decomposition), and E is exudation of dissolved organic carbon (DOC) from phytoplankton. Sedimentation of phytoplankton and phytodetritus is high immediately after the spring and autumn blooms and low during summer (Turner, 2002). Phytoplankton exudation of DOC is considered to be around 5–10% of net productivity (Banse, 1995). Therefore, during summer, as S is low, $\Delta A \sim 0$ and E constitute a small fraction of P, grazing, G, must be high to account for the high primary productivity. Conversely, during spring and autumn blooms, grazing must be relatively low.

Nutrients from zooplankton exudation and excretion are reutilized by phytoplankton (Sterner, 1986). Hence, during spring/summer, new nutrients (allochthonous) from deeper water are captured and recycled in the plankton community. When the import of new nutrients has come to a halt around July (Fig. 6.2b), primary production, including the autumn bloom, is based on recycled nutrients (autochthonous). Consequently, nutrients change from new nutrients in spring to the dominance of recycled nutrients in summer.

There are two pathways for nutrient cycling, via the microbial loop and via the grazing food web, with intermediate stages between the two (Legendre and Rassoulzadegan, 1995). Small phytoplankton (less than 5 μm) are the main primary producers, and microzooplankton (heterotrophic flagellates and ciliates ranging from 20 to 200 μm) are the main grazers in the microbial loop. Copepods are the main herbivores in the grazing food web (Fenchel, 1988). Dilution experiments suggest that microzooplankton are the main consumers of primary production across a wide range of marine ecosystems (Calbet and Landry, 2004). However, there is high variability between the various experiments. It has also been argued that dilution experiments tend to overestimate microzooplankton grazing (Dolan and McKeon, 2005). In the Skagerrak–North Sea area copepods show a seasonal pattern similar to that of primary productivity (Fig. 6.2a) with peak abundance around July (Fransz et al., 1991; Kiørboe and Nielsen, 1994). In contrast, microzooplankton seem most abundant during spring and autumn blooms (Nielsen and Kiørboe, 1994). This suggests that the grazing food chain constitutes a relatively important pathway for nutrient cycling during summer in these waters.

Numerous studies have shown that copepods have a highly developed selective feeding mechanism that can discriminate on the basis of size (Wilson, 1973), taste (DeMott, 1988; Kerfoot and Kirk, 1989), food quality (DeMott, 1989), and toxicity (Turner and Tester, 1997). Similar selective feeding mechanisms have been observed in micozooplankton grazers too (Verity, 1991; Buskey, 1997). With this in mind, Kerfoot and Kirk (1989) pointed out that it is easy to conceive of the evolution of unpalatability in phytoplankton as a defense against grazers. About 7% of the total estimated number of phytoplankton species have been reported to produce red tides, and 2% are harmful because of their biotoxins, anoxia, irradiance reduction, or nutritional unsuitability or because they cause physical damage and so forth (Smayda, 1997b).

Due to an excess of nutrients in spring, larger diatoms escape grazing control and bloom due to the low biomass of over-wintering stocks of mesozooplankton (Riegman et al., 1993; but see Chapter 8). After the spring bloom, grazers generally control the phytoplankton biomass. If grazing rates had been higher or lower than algal growth rates, this would have led to changes in the algal biomass over time. As the algal biomass remains quite stable during summer (Fig. 6.2a), the average grazing rate must balance growth rates in phytoplankton. This is in agreement with the general perception of control in marine plankton communities, as most clearly stated by Legendre (1990, p. 689): "More generally, large phytoplankton outbursts are simply not possible when grazing pressure is high." Interestingly, this implies dominance of grazed phytoplankton under high grazing pressure and leads to two important questions: Why do low- and non-grazed phytoplankton not take advantage of the high grazing pressure and bloom, and how can dominance of grazed phytoplankton under high grazing pressure be an evolutionary stable strategy (ESS), that is, it cannot be invaded by a mutation resulting in non-palatability in phytoplankton?

There are two possible solutions why there is grazing control of the phytoplankton community in summer: (1) all phytoplankton species are grazed at the rates by which they grow (there are no inedible phytoplankton), or (2) edible phytoplankton species have competitive advantages over low- and non-grazed species.

Theoretically, all phytoplankton species could be grazed at rates that balance growth, for example, by the grazers "killing the winners" (Thingstad, 1998). For this to be an ESS there must be mechanisms that prevent evolution of non-palatability in phytoplankton. One possible explanation could have been that the evolutionary arms race results in such a balance, that is, a mutation in an alga that result in reduced grazing is quickly balanced by improved capabilities in the grazers. However, the autumn bloom that consists of mainly red tide forming dinoflagellates (Chapter 8; Dahl and Johannessen, 1998), *Karenia mikimotoi*, which is toxic (Yasumoto et al., 1990), and *Ceratium* spp., which are large and therefore not efficiently grazed (Granéli et al., 1989), supports neither the perception that all phytoplankton are grazed at the same rates as they grow nor that large phytoplankton outbursts are impossible when grazing pressure is high (Legendre, 1990). The autumn outburst occurs when primary productivity is close to maximum and algal biomass is low, which, according to the conceptual model for primary productivity above (Eq. 6.1), imply high grazing rates. Furthermore, there is no evidence of new nutrients that trigger the autumn bloom or that the phytoplankton utilize nutrients at deeper water by vertical migration (Fig. 6.3). Also, one would expect the benefits of vertical migration to be apparent throughout summer, as the spring bloom depletes allochthonous nutrients in the surface layer. The onset of the autumn bloom concurs with decreasing primary productivity and nutrients being lost from the euphotic layer (Figs. 6.2 and 6.3). However, primary productivity is still high at the autumn outburst, which implies ample nutrients for algal growth. Johannessen et al. (2006) suggested that the autumn bloom could be explained in terms of overgrazing that prevents edible algae from fully utilizing recycled nutrients, which therefore become accessible for low- and non-grazed species (more details follow). Therefore, as low- or non-palatability seems to be a successful strategy under certain trophic or environmental conditions, I conclude that the proposed solution—that all phytoplankton are edible and therefore grazed at the same rates as they grow—is inadequate. This leads to the second solution, namely that edible phytoplankton species have competitive advantages over low- and non-grazed species.

6.4.3 Predator-Prey Synergism

The abundance of phytoplankton is determined solely by growth and mortality. During summer, average algal growth rates in Skagerrak balance grazing. As low- and non-grazed species per definition experience low mortality, their net growth rates (at the population level) must be very low for them not to bloom. For example, moderate net growth rates of $\mu = 0.2\ d^{-1}$ and $\mu = 0.3\ d^{-1}$ will result in 15- and 51-fold increases in abundance in 15 days, respectively. Growth in phytoplankton depends on light and nutrients. Light is available for all phytoplankton. Consequently, nutrients seem to be the only factor that may provide edible phytoplankton with competitive benefits over low- and non-grazed algae. Because herbivorous zooplankton recycle a substantial fraction of nutrients of the ingested food (Lehman, 1980; Sterner, 1986) and alter the nutrient composition (Andersen and Hessen, 1991; Elser and Urabe, 1999), it seems likely that the benefits for edible phytoplankton involve a positive feedback from nutrient cycling. At microscales there is evidence of higher concentrations of nutrients from zooplankton exudation and excretion (Blackburn et al., 1998; Lehman and Scavia, 1982; Seymour et al., 2009), and high uptake rates coupled with the ability to store nutrients enable the phytoplankton to maintain nearly maximum rates of growth at media concentrations that cannot be quantified with traditional analytical techniques (McCarthy and Goldman, 1979). Accordingly, high primary productivity coupled with low algal biomass in summer (Fig. 6.2a) implies that algal growth rates as well as nutrients available for algal growth are high.

To develop a mechanistic explanation for the benefits of edible phytoplankton during high grazing pressure in summer, we should approach the problem from the scale at which the biological interactions take place, namely at microscales. In addition to accounting for the advantage for edible phytoplankton, the mechanism should indicate how this advantage could be an ESS. Fig. 6.4 suggests such a mechanism.

Consider a sparse algal community (stage 1) consisting of a preferred algal prey species, a, with different genotypes, $a_{1\text{-}5}$, an edible but

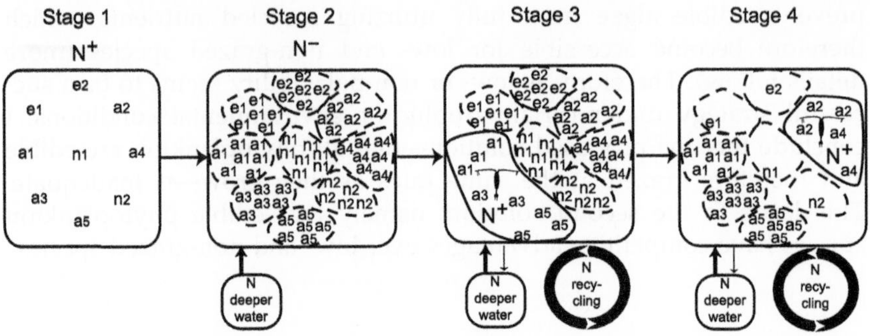

FIGURE 6.4 A conceptual model illustrating the mechanism as to why being eaten is advantageous in phytoplankton, in terms of four successive temporal stages. N^+—excess of nutrients, N^-—depleted nutrients, a–a_5–five different genotypes of a preferred algal prey species, e_1 and e_2—two different genotypes of an edible but not preferred algal species, n_1 and n_2— two different genotypes of a non-edible algal species, T—herbivore.

non-preferred species, $e_{1\text{-}2}$, and a non-edible species, $n_{1\text{-}2}$. If allowed to grow in excess of nutrients (N^+), the community will become spatially structured at microscales because each algal cell will give rise to a patch of daughter cells due to asexual reproduction (stage 2). Without grazing, nutrients will become exhausted (N^-). Herbivores (micro- or mesozooplankton) will search for high concentration of their preferred algal prey (stage 3), and while grazing simultaneously exude nutrients (Lehman and Scavia, 1982; nutrient cycling is described in Fig. 6.5). In order to optimize foraging, herbivores will move on to higher cell concentrations before having grazed all cells, allowing the remaining algae to increase their abundance again (stage 4). Scavia et al. (1984) observed that a substantial part of the algal cells entrained in the feeding currents of zooplankton were released unharmed. The main source of nutrients during summer stems from recycling, and recycling appears to takes place at microscales (Blackburn et al., 1998; Lehman and Scavia, 1982; Seymour et al., 2009), so patches of non-grazed and non-preferred cells will receive substantially less nutrients, become nutrient limited, senescent, and appear in low concentrations (stage 4; the following objections are discussed in relation to nutrient cycling). It should be noted that the suffixes indicating different genotypes could also represent different species as long as the phenotypical traits are the same (e.g., $a_{1\text{-}5}$ could represent five different species of preferred algal prey and n_{1-2} two different non-edible species).

As the herbivores retain some of the ingested nutrients, a corresponding input of new nutrients has to be supplied for the system to be in balance. These nutrients are supplied from deeper water, as evidenced by nitrate being utilized gradually deeper down the water column (Fig. 6.2b). After the spring bloom, primary productivity continues

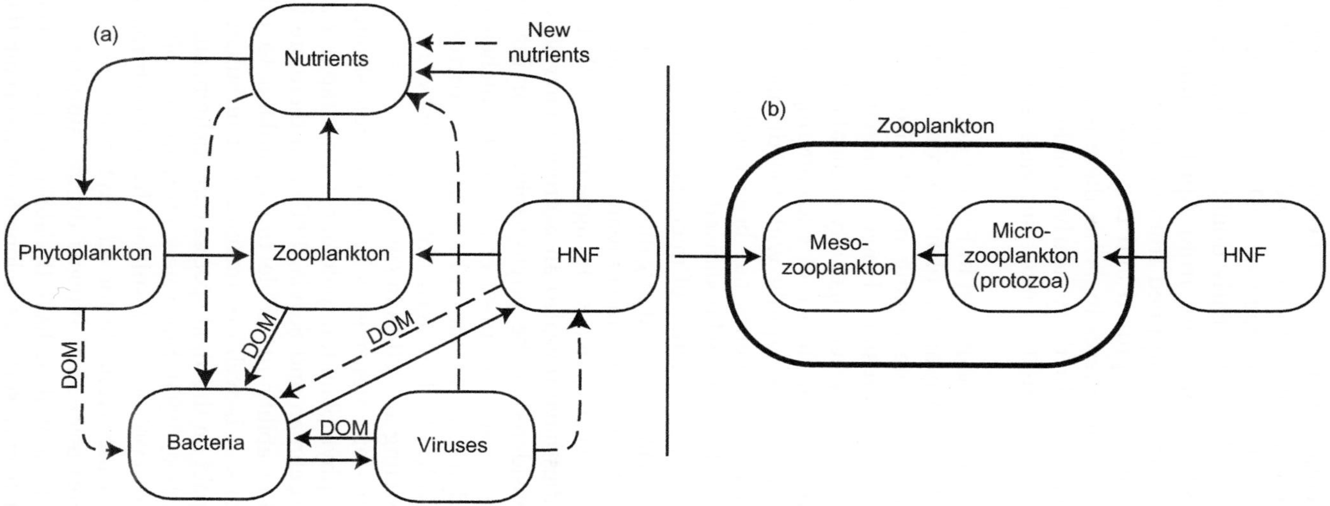

FIGURE 6.5 Two potential pathways of nutrient recycling (see text for explanation).

to increase until June (Fig. 6.2a). A part of this increase can be attributed to increased day length. However, the measured light factor (LF), which was used to estimate daily primary productivity ($P/d = P/h \times LF$), increased merely from an average of 10.0 in April to 11.5 in June, whereas primary productivity nearly doubled. Therefore, most of the increase in primary productivity from April to June is probably a result of nutrients that are supplied from deeper water to the euphotic layer, where they are captured and recycled, and to a less extent to increased day length. Interestingly, as there is no increase in algal biomass and primary productivity nearly doubles from April to June, growth rates in phytoplankton must increase substantially during this period.

It should be noted that the model in Fig. 6.4 does not pretend to provide a fully realistic picture of the plankton community. For example, there will probably be some mixing of cells from different patches (swimming, turbulence, diffusion). However, asexual reproduction will inevitably result in spatially heterogeneity of algal cells both in terms of genotypes and species (the chance of encountering clonal copies of a specific algal cell decreases rapidly with the distance from the cell). Such spatial structuring of genes has been confirmed in a modeling study by Young et al. (2001), who showed that phytoplankton become intermittently distributed in terms of micropatches merely as a result of the fundamental asymmetry that organisms always are born in company but may die alone. "Born in company" implies that there will be high relatedness in such patches. In agreement with this modeling study, in situ studies have revealed an isotropic starry view of fluorescence patches considered to be phytoplankton, varying in size from millimeters to centimeters (Doubell et al., 2009; Mitchell et al., 2008). Hence, the underlying assumption of a spatially structured algal community at microscales, both in terms of genes and clusters of cells, is confirmed by modeling as well as in situ studies.

Turbulent mixing of water masses by for example strong winds, may lead to reduced phytoplankton patchiness and rapid dissipation of nutrient patches (Lehman and Scavia, 1982) and therefore have a negative impact on the ability of edible phytoplankton to utilize micropatches of nutrients. It is well documented that turbulent mixing does have a strong impact on the algal species composition, such as by favoring diatoms (Kiørboe, 1993). Since new nutrients are often mixed into the euphotic layer in pulses as a result of physical disturbances, such as strong winds, and are available for all phytoplankton, mixing events may result in short-term algal blooms during summer. On the other hand, nitrogen-sufficient algae have the ability to accumulate intracellular nitrogen, which may serve as a long-term reservoir (Dortch et al., 1984). Hence, grazed algae may sequester a substantial part of the new nutrients if they dominate prior to the mixing event, as they are expected to do according to the model in Fig. 6.4.

Obviously, the grazers will benefit from stimulating growth of their preferred algal prey. Hence, this is win-win situation for phytoplankton and their grazers. I call such a positive relationship between trophic levels predator-prey synergism—the abundance of both predator and prey will be enhanced by their coexistence. In contrast to mutualism, predator-prey synergism gives no obvious benefit at the individual level because the herbivore predator kills its algal prey. It should be noted though that the proposed mechanism of positive coexistence of phytoplankton is in agreement with the perception of "the selfish gene" as remnant genetically identical cells continue to grow and multiply.

6.4.4 Nutrient Cycling

The importance of microbial processes in the sea has become increasingly apparent over the past three decades (Fenchel, 2008; Fuhrman, 1999), and heterotrophic bacteria and protozoans, mainly heterotrophic nanoflagellates (HNF), have been identified as key players in the cycling of nutrients (Azam et al., 1983). Fig. 6.5 depicts the cycling of nutrients when herbivores control the phytoplankton community (the summer situation). The model is modified from Legendre and Rassoulzadegan (1995) and Fuhrman (1999). Phytoplankton are grazed by zooplankton. Dissolved organic matter (DOM) from zooplankton exudation, excretion, and sloppy feeding is consumed by bacteria. A smaller proportion of DOM consumed by bacteria stems from phytoplankton exudation (indicated by a broken line; Banse, 1995). Bacteria are grazed by HNF (Sanders et al., 1992). As prokaryotes (lacking cell nucleus) have a higher nitrogen and phosphorus concentration per volume of biomass than eukaryotes, HNF grazing bacteria release excess nutrients that are not required for their growth (Pernthaler, 2005). HNF, in turn, are grazed by larger zooplankton, varying in size from small protozoans such as ciliates and rotifers to larger mesozooplankton like copepods and cladocerans (Jürgens et al., 1996). However, as there is generally a positive relationship between predator and prey size in plankton (Hansen et al., 1994), it is conceivable that in the case of larger zooplankton grazers, HNF may be controlled by microzooplankton, such as ciliates, which in turn are controlled by the larger grazers (Fig. 6.5b).

Viruses are ubiquitous and the most abundant biological agents in the sea, typically appearing in numbers of $\sim 10^7$ viruses ml^{-1} (Fuhrman, 1999). They are positively related to bacterial abundances, suggesting that the majority of viruses are phages (Weinbauer, 2004). Viral-associated mortality is highly density-dependent on the density of bacteria. Hence, phages will primarily affect the largest and fastest growing bacterial populations (Pernthaler, 2005), a phenomenon known as "killing the winners" (Thingstad, 2000). In marine systems it is assumed that 20–40% of the bacteria are killed by viruses on a daily basis (Suttle, 2005). Viruses are transforming bacterial biomass to DOM, which is reutilized by bacteria (Middelboe and Jørgensen, 2006), thus forming a semi-closed loop (Fuhrman, 1999). A variable proportion of viruses are grazed by HNF, but other factors seem to be more important for viral decay (Bongiorni et al., 2005).

Although cycling of nutrients is much more complicated than indicated in Fig. 6.4, this is not in conflict with the idea of a positive relationship between zooplankton grazers and their preferred algal prey. Because bacteria and HNF reproduce asexually, genes will become spatially structured at microscales, as in phytoplankton (the model by Young et al., 2001, applies to all organisms reproducing by binary fission). By contributing to cycling of nutrients, both bacteria and HNF may gain competitive advantages by sacrificing parts of their clonal populations in order to obtain resources for continuous growth. As ciliates generally reproduce asexually, the argument of the benefit of sacrificing clonal copies is applicable in the case of a larger mesozooplankton grazer, too (Fig. 6.5b). Consequently, it can be argued that all organisms taking part in the recycling process, and that reproduce by binary fission, benefit from coexisting with their grazers, by the cost of losing cells being outweighed by the benefits of receiving resources for continuous growth. If growth does not balance grazing, for example with bacteria being undergrazed, nutrients will become entrapped in the increasing bacterial biomass and nutrient cycling impeded. Interestingly, viruses may act as predators on "the winners" and thereby contribute as positive players in the recycling process. This may explain why viruses are highly abundant and ubiquitous in aquatic environments.

Lehman and Scavia (1982) demonstrated in a laboratory experiment that phytoplankton can sequester nutrients from ephemeral plumes from zooplankton exudation. This was questioned by Currie (1984), who showed in a modeling study that in nature such patches might dissipate by diffusion too quickly to be used by the phytoplankton. However, the modeling exercise by Currie was based on the assumption that phytoplankton are homogeneously distributed, and the model did not include swimming behavior of the phytoplankton. In a classical paper, Berg and Purcell (1977, p. 205) concluded that in a "uniform medium motility [swimming or sinking] cannot significantly increase the cell's acquisition of material." This applies to cells less than 10 μm and is a result of the high surface tension and high viscosity at such small scales (Mann and Lazier, 2006). In a non-uniform medium, on the other hand, a motile organism can seek out more favorable regions using chemotaxis (Berg and Purcell, 1977). Chemotaxis has been observed in both phytoplankton

(Govorunova and Sineshchekov, 2005; Seymour et al., 2010) and bacteria (Blackburn et al., 1998; Stocker et al., 2008). In an elegant laboratory study, Seymour et al. (2009) showed that a phytoplankton, a heterotrophic bacterium, and a phagotrophic flagellate were all able to exploit microscale patches of their resources, and the authors pointed out that microscale nutrient patchiness may subsequently trigger the sequential formation of patches of phytoplankton, heterotrophic bacteria, and protozoan predators in the ocean. Hence, motile aquatic microorganisms have the ability as well as the capacity to exploit ephemeral microscale resources, a quality that will provide them with substantial competitive benefits when resources are intermittently distributed at such scales.

6.4.5 Evolutionary Considerations

The models in Figs. 6.4 and 6.5 outline the principle behind the proposed positive coexistence of zooplankton and their preferred algal prey and provide a solution to the question of why grazed phytoplankton dominate over little- and non-grazed species under high grazing pressure. One important question still remains: how can this be an evolutionary stable strategy?

A mutation resulting in a non-palatability phenotype that has all the other abilities of its palatable conspecifics would initially have a higher growth rate because of reduced grazing mortality. However, as it continues to grow, the concentration of non-palatable cells will increase, as well as the space they occupy. Grazers will eventually not visit such patches, and nutrient cycling therefore will become severely impeded. The non-palatable phenotype will no longer benefit from the cycling of nutrients and therefore have lower fitness than its palatable conspecifics. Accordingly, it can be argued that mutations resulting in non-palatability in bacteria or HNF (and protozoan microzooplankton, Fig. 6.5b) will disrupt nutrient cycling, as nutrients will become entrapped in the increasing biomass of the non-palatable phenotypes. Hence, I conclude that the advantage in phytoplankton, bacteria, and HNF of sacrificing parts of their clonal populations in order to obtain resources for continuous growth is an ESS.

The proposed mechanism is based on asexual reproduction in phytoplankton. However, from a theoretical point of view, the mechanism could also work in sexually reproducing organisms due to spatially heterogeneity of genes: the closer together, the higher the similarities in genes. This allows for kin selection and evolution of altruism.

In combination with the microscale recycling hypothesis (Fig. 6.4), one can speculate about additional mechanisms for the beneficial coevolution of phytoplankton and herbivores: the stoichiometry (composition) of recycled nutrients (Andersen and Hessen, 1991; Elser and Urabe, 1999), zooplankton selectively excreting fecal pellets in patches containing preferred phytoplankton (nutrients dissipate readily from fecal pellets, Jumars et al., 1989), and zooplankton exerting a direct negative impact on non-grazed species (e.g., by bolus rejection [Scavia et al., 1984] or physical destruction).

6.4.6 Examples of Altruism in Clonal Populations

In an evolutionary perspective, sacrificing part of the clonal population to gain competitive advantages can be considered equivalent to that of producing public goods, which is a relatively common form of social behavior in microbes (West et al., 2007). Public goods are products manufactured by an individual (e.g., a bacterium) that can be utilized by the individual and its neighbors, clonal copies but also cheaters that do not produce public goods. Although there is metabolic cost of producing public goods, bacteria with such phenotypical traits have been shown to outcompete selfish mutants under conditions of relatively high relatedness in the bacteria (Griffin et al., 2004). Cheaters, on the other hand, do better in lower frequencies in the population because they are then better able to exploit the co-operators (West et al., 2007, and references therein). High relatedness in intermittently distributed micro patches is indeed what is expected in phytoplankton and other marine microbes reproducing by binary fission (Young et al., 2001).

Another example of altruistic behavior in microbes is abortive infection (Abi) systems, also called phage exclusion, which are found in many bacterial species (Chopin et al., 2005). Abi are genetically activated mechanisms that cause premature bacterial cell death following phage infection (Forde and Fitzgerald, 1999). This prevents phage multiplication and limits the spread of viral progeny particles to other bacterial cells, thereby allowing the bacterial population to survive. (It is questionable whether sacrificing part of a clonal population to survive or to gain competitive advantages is true altruism when the survivors are genetically identical.)

Programmed cell death (PCD) has been observed in a number of unicellular phytoplankton species when subjected to environmental stresses, such as nutrient deprivation or low light (Bidle and Falkowski, 2004, and references therein). Bidle and Falkowski proposed that PCD may have evolved in phytoplankton cells as a strategy to relieve a population of nutrient stress and remove ageing and/or damaged cells, but the authors stressed that the release of nutrients into the water by PCD would seem to likely benefit both competing species and closely related or clonal individuals. However, if the nutrients released from PCD would be sequestered at microscales by clonal copies, there are no evolutionary obstacles against the hypothesis proposed by Bidle and

Falkowski (2004). In principle, PCD would be in agreement with the mechanism proposed here that phytoplankton, bacteria, and HNF gain competitive advantages by sacrificing parts of their clonal populations in order to obtain resources for continuous growth.

Rynearson and Armbrust (2005) estimated the total number of clonal lineages in a spring bloom of *Ditylum brightwellii* to be approximately 2500. With up to 10,000 cells l^{-1} the number of genetically identical cells in the population must have been formidable. Therefore, if all clonal cells in a micropatch of *D. brightwellii* should become grazed, this would not pose a threat to the clonal lineage. In agreement with this, Rynearson and Armbrust (2004, 2005) identified the genetic signature of the bloom population of *D. brightwellii* several times during a period of three years and took this as an indication that populations as well as individual clonal lineages are long-lived and able to survive environmental variation.

6.4.7 Autumn Blooms and Red Tides

At the onset of the autumn bloom, nitrate is exhausted in the upper layer, whereas, below 30 m depth, nutrients commence increasing (Fig. 6.2), probably due to sedimentation outweighing processes bringing nitrate to the surface. Furthermore, precipitation does not increase notably from July to August and September (Johannessen et al., 2006). Hence, there is no evidence of new nutrient supply triggering the autumn outburst. Rather, reduced primary productivity coupled with increasing nutrient concentrations in deeper water suggest that nutrients are being lost from the euphotic zone at the commencement of the autumn outburst. On the other hand, as primary productivity is still high, the nutrient supply available for phytoplankton growth must be adequate. As there is no evidence of extra allochthonous nutrients being available, primary production must be based on recycled nutrients from zooplankton exudation and excretion.

Johannessen et al. (2006) suggested that there are three possible mechanisms that could deprive edible algae of their competitive advantages and lead to the autumn outburst of dinoflagellates: (1) a lack of grazing control because of reduced herbivorous stocks (zooplankton going into diapause), (2) unfavorable environmental conditions (e.g., turbulent mixing), or (3) zooplankton overgrazing their preferred algal prey. The problem with mechanisms 1 and 2 is that neither reduced herbivorous stocks nor unfavorable environmental conditions can explain why little- and non-grazed algae dominate the autumn bloom. Both of these conditions imply a lack of grazing control, and therefore the species composition of the autumn bloom is determined purely by physical environmental conditions. Such conditions should favor fast-growing species and not the large and slow-growing *Ceratium* (Weiler and Eppley, 1979).

Therefore, a mechanism that prevents edible and fast-growing algae from blooming needs to be identified, which leads to explanation 3, overgrazing of palatable algal species. Overgrazing will prevent edible algae from fully utilizing recycled nutrients, which will become available for little- and non-grazed species. Because the commencement of the autumn bloom coincides with declining primary productivity, Johannessen et al. (2006) suggested that an explanation for overgrazing could be a delay in the decrease in herbivorous biomass relative to primary productivity, which would subsequently cause an imbalance between herbivores and palatable algae. In principal, synergistic herbivores overgrazing their algal prey is similar to the impact antagonistic herbivores will have on the phytoplankton community. In conclusion, the autumn bloom of little grazed dinoflagellates appears to be a result of trophic processes rather than environmental conditions.

Lindahl and Hernroth (1983) reported that the autumn bloom of *Ceratium* spp. and *K. mikimotoi* started to appear around the mid-1970s. After 2002, the autumn bloom had practically vanished (Chapter 5; Johannessen et al., 2012). Interestingly, both the appearance and disappearance of the autumn bloom coincided with abrupt changes in the plankton community (Chapter 5). This supports the perception that the autumn outburst of dinoflagellates is a result of trophic processes.

Mitra and Flynn (2006) have proposed a similar mechanism in which harmful algal blooms (HAB) can develop through the self-propagating failure of normal predator–prey activity, resulting in the transfer of nutrients into HAB growth at the expense of competing algal species. They suggested that under nutrient limitation a phytoplankton species that is initially palatable might become less palatable and bloom at the expense of a more palatable species that is overgrazed. However, this mechanism is unlikely to be relevant for the success of *Ceratium* spp. because large size is considered to be the reason why these algae are not efficiently grazed (Granéli et al., 1989). The mechanism proposed here does not depend on shifts in the strategy in phytoplankton but suggests that overgrazing of edible algae promotes growth of non-palatable algae through increased availability of recycled nutrients.

Red tides may occur if the water column remains highly stratified for several weeks during summer (Margalef et al., 1979). Smayda (2002, p. 95) points out that "there is a contradiction in the stratification-HAB paradigm: the conditions of a stratified water-mass, with its characteristic nutrient-poor conditions, are expected to promote stasis of the population, rather than growth and blooms." Calm conditions during summer with a highly stratified water column hamper mixing and the import of nutrients from deeper water, while nutrients may be lost from the euphotic zone due to sedimentation. Hence, calm conditions during summer may result in similar conditions as during the autumn outburst, with reduced primary productivity, zooplankton overgrazing edible algae, and the proliferation of little- and non-grazed species at the expense of palatable phytoplankton. Consequently, the non-palatable strategy can be regarded as an opportunistic strategy in which phytoplankton take advantage of the collapse of synergism.

From the model in Fig. 6.4, it can be deduced that in eutrophic environments where there are excess of allochotonous (new) nutrients in summer, the non-palatable strategy of phytoplankton may also be successful and result in high algal biomass. Consequently, overgrazing as well as undergrazing (i.e., there are more nutrients than the herbivore-phytoplankton synergists can utilize) may result in the promotion of little- and non-grazed phytoplankton (growth rates in phytoplankton must balance grazing for synergism to operate). This is important to keep in mind when evaluating the synergism hypothesis against published marine and limnic studies. Strong doses and thereby imbalance between trophic levels seem to be quite common in experimental studies (see Chapter 7 for more detailed discussion). Furthermore, microscale grazing studies using continuous or intermittent stirring keeping the phytoplankton suspended will break down the microscale structures in the water column and render synergism impossible.

6.4.8 Strategies in Phytoplankton

Different strategies in the phytoplankton can be envisaged from the seasonal succession (Fig. 6.2a). In the turbulent and nutrient-rich environment of spring, larger diatoms escape grazing control and bloom due to the low biomass of over-wintering stocks of mesozooplankton (Riegman et al., 1993). After the spring bloom, zooplankton biomass increases and the phytoplankton biomass becomes controlled by the grazers. In autumn, little- and non-grazed species bloom. This leads to three main strategies in the phytoplankton: (1) The rapid growth-turbulence strategy—phytoplankton outgrow grazers in turbulent and/or nutrient-rich environments; (2) the synergistic strategy—phytoplankton coexist in balance with their grazers; (3) the non-palatable strategy—an opportunistic strategy where the phytoplankton take advantage of the collapse of synergism.

6.4.9 Ecosystem Resilience and Bifurcations

The theoretical relationship between resilience, environmental changes, perturbations, and bifurcations is presented in Fig. 1.1. The idea is that under optimal environmental conditions for a particular state, the resilience is high and perturbations are unlikely to cause shifts to alternative states. If the environmental conditions change gradually in favor of another community structure, the resilience is reduced and the system rendered vulnerable to shift abruptly to the alternative state from perturbations the system under optimal environmental conditions could withstand. Perturbations may, for example, be in the form of mass mortality caused by diseases, toxic algal blooms, or low or high temperatures, which may pave the way for organisms that are better adapted to the altered environmental regime. Hence, under this interpretation of the mechanism underlying bifurcations, the organisms with competitive advantages will dominate the system after a severe perturbation.

The repeated incidents of abrupt and persistent recruitment failures in gadoids along the Norwegian Skagerrak coast (Chapters 2 and 5), which were linked to sudden changes in the plankton community as a result of gradual environmental changes (eutrophication and temperature, Chapters 3, 4, and 5), suggest that marine ecosystems are indeed vulnerable to shift between alternative stable states. As these shifts probably occurred in the plankton community, which, in turn, were propagated to higher trophic levels by causing recruitment failure in fishes, there must be mechanisms within the plankton community that prevent the system from responding in a gradual dose-response manner, that is, resilience. Interestingly, synergistic interactions in plankton may provide a mechanistic explanation for resilience.

Theoretically, interaction between herbivorous zooplankton and their algal prey may be either positive or negative. There are two major advantages to synergism when compared to antagonism in grazers. First, by stimulating the production of their preferred algal prey, synergistic grazers allocate maximal energy to themselves, whereas negatively interacting grazers will reduce the abundance of their prey and provide competing algal species and their grazers with benefits. Second, synergists are able to dampen the impact of suboptimal environmental conditions, whereas negatively interacting grazers will exacerbate unfavorable conditions for their algal prey.

Because antagonistic herbivores have a negative impact on the abundance of their algal prey, they depend on favorable environmental conditions for their algal prey to become abundant themselves and to remain abundant. Organisms for which at any time the environmental conditions are optimal will thus dominate, and the plankton community therefore behaves in a simple dose-response manner to fluctuating and changing environmental conditions. Furthermore, as antagonistic herbivores will reduce the abundance of their algal prey and thereby pave the way for competing algal species, there is no mechanism to prevent invasion of species from different community structures to flourish under (for them) optimal environmental conditions. Consequently, negative interaction between herbivores and their algal prey appears incompatible with high resilience and abrupt shifts in the plankton community.

In contrast, by stimulating growth of their preferred algal prey, synergistic herbivores will have a strong and positive influence on the algal community and, at least to some extent, override bottom-up processes, thereby being able to persist under suboptimal environmental conditions and prevent species from other community structures to invade and establish. Consequently, synergism in plankton may provide a plausible explanation for resilience in the plankton community.

6.4.10 Predator-Prey Synergism in Terrestrial Ecosystems

Many people who live in the countryside may observe apparently positive predator-prey relationships practically on their doorstep. Fig. 6.6, a photograph taken close to my cabin in the Norwegian mountains, illustrates this phenomenon. In Norway, farmers take their sheep to the mountains to graze in summer. In the early 1960s the area in the picture was a dumping ground for rock masses, which were subsequently covered with soil and sown with grass. Since then the area has remained pastureland. In the 10 to 15 years before the picture was taken,

FIGURE 6.6 Grassland in the Norwegian mountains. Thicket of trees developed inside an enclosure where sheep were prevented from grazing.

farmers built enclosures in which to collect their sheep in the autumn. During summer, these enclosures were fenced and could no longer be grazed by sheep. Within a few years the enclosed land developed into a thicket of trees, impenetrable to sheep and too dark for grass. Outside the enclosures, grass continued to dominate, implying that grass and other pasture plants gain competitive advantage from sheep grazing, and the sheep benefit from "farming their environment." Furthermore, the primary productivity of the grass may increase with grazing, owing to the recycling of nutrients (McNaughton et al., 1997). Sheep probably graze tree seedlings, thus preventing them from outcompeting pasture flora. Broom and Arnold (1986) observed that in a sheep pasture, capeweed plants (*Arctotheca calendula*) became less abundant despite not being grazed by sheep. They noticed that when sheep grasped the capeweed plants they sometimes pulled them up by the roots and then dropped them. Hence, there is evidence that sheep weed their pasture.

There are similar examples from natural ecosystems. In Lake Manyara game park in Tanzania, dramatic reduction of impala numbers (*Aepyceros melampus*) following outbreaks of rinderpest and anthrax enhanced seedling recruitment of *Acacia* trees. This resulted in even-aged stands of *Acacia* and substantial bush encroachment of the game park. Prins and van der Jeugd (1993) suggested that seedling establishment of *Acacia* is rare due to ungulates grazing seedlings, and that epidemic disturbances among ungulates create narrow windows for seedling recruitment and thus even-aged stands of *Acacia*.

Savanna ecosystems are characterized by the co-dominance of trees and grasses. Sankaran et al. (2005) showed that maximum woody cover in savannas receiving a mean annual precipitation (MAP) of less than 650 mm increases linearly with MAP, thus allowing grasses and trees to coexist. Above 650 mm, MAP is sufficient for woody canopy closure, and disturbances in the form of fires and herbivory are required for the coexistence of grasses and trees. Also below 650 mm, MAP woody cover is held below maximum by fires and herbivory. Grassland ecosystems are one of Earth's dominant biomes (Strömberg, 2011). In these systems coexistence of grazers and grasses appears to induce increased abundances of both trophic levels and can thus be considered synergistic.

In grassland systems in which the grazers are important to control the abundance of tall vegetation, reduced herbivorous stocks may result in loss of habitat for the grazers, as in Lake Manyara. However, overgrazing may result in loss of habitat, either by desertification (e.g., Sinclair and Fryxell, 1985) or shrub encroachment (e.g., Roques et al., 2001). Hence, it is tempting to speculate whether carnivores predating herbivores might also be a synergistic strategy by the carnivores controlling the abundance of the herbivores. It seems reasonable to assume that herbivores overgrazing will affect all three trophic levels negatively.

Many vegetation types depend on fire, and Mutch (1970, p. 1047) hypothesized that "fire-dependent plant communities burn more readily than non-fire-dependent communities because natural selection has favored development of characteristics that make them more flammable." The hypothesis has been criticized for being group-selectionist (Snyder, 1984). However, Bond and Midgley (1995) modeled the evolution of flammability and concluded that flammability may enhance inclusive fitness if the resulting fires kill neighboring, less-flammable individuals and also open recruitment possibilities. Hence, under such conditions fires can be considered equivalent to herbivory (Bond and Keeley, 2005). Flammable vegetation will increase in abundance with fires, which in turn will increase in frequency and intensity in flammable system. Hence, flammability may be considered a synergistic strategy.

The black-tailed prairie dog (*Cyonomus leucurus*) has been observed to behave in ways that have been equated with farming their environment (Gordon and Lindsay, 1990, and citations therein), and McNaughton (1984) argued that coevolution among plants and animals has resulted in animals utilizing their food resources in such a way as to increase the yield of their resources. Owen-Smith (1987) proposed that the demise of about half the mammalian genera exceeding 5 kg in body mass in the late Pleistocene might have been caused by changes in the vegetation as a result of human hunters eliminating megaherbivores (greater than 1000 kg).

The idea of mutualism between plants and herbivores is not new (e.g., Owen and Wiegert, 1981), but the discussion around this controversial topic has mainly revolved around the question as to how herbivory may increase fitness of a plant relative to a non-grazed specimen of the same species (Bergelson and Crawley, 1992; Paige and Whitham, 1987), that is, classical mutualism. However, the beneficial coexistence of predator and prey does not operate through a direct increase in individual fitness of the prey but rather through competitive benefits outweighing the costs of predation. There are potentially some evolutionary problems in relation to this perspective that need to be investigated.

Like most biologists, I am not comfortable with the idea of group selection (Wynne-Edwards, 1986), that is, adaptations that benefit the group but not the individual (gene). One can easily perceive how palatability in grasses may evolve, but it is less obvious how palatability can be an evolutionary stable strategy in the face of mutation of non-palatability. Grassland ecosystems have evolved and expanded since the Late Cretaceous and are today a dominant biome (Strömberg, 2011). Hence, it appears likely that there are mechanisms that prevent mutation of non-palatability to spread. One explanation could be that the grazers provide palatable specimens with higher fitness by affecting non-palatable specimens even more negatively (grazer control), as with sheep weeding out capeweed (Broom and Arnold, 1986). Also, it is possible to conceive of a mechanism that is independent of the behavior of the grazer. If a fitness-enhancing mutation of non-palatability should occur, the mutant will increase in the pasture. As the mutant gradually replaces palatable pasture plants, it will no longer benefit from grazing and eventually become replaced by taller vegetation. If the mutant could spread to the entire population, the species would become extinct. On the other hand, if the extinction appears locally, the palatable phenotype could again spread. As non-grazed small pasture plants indirectly depend on grazing, the same mechanisms (grazer control and local extinctions) that prevent successful mutation of non-palatability in grass may operate to optimize abundance of non-grazed plants. Hence, there appear to be mechanisms that prevent mutation of non-palatability to be successful, and thus palatability in grasses to be an evolutionary stable strategy. Furthermore, the evolution of palatability does not rely on group selection because all individuals of a species will be palatable, grazed, and therefore their competitive abilities equally affected. The gain in fitness of grazing is thus not related to intra-specific competition or inter-specific competition with pasture plants but competition with taller vegetation. There are probably other processes that determine the diversity of the pasture by affecting intra-specific and inter-specific competition in pasture plants.

In conclusion, is appears that predator-prey synergism is important in some terrestrial ecosystems.

6.5 CONCLUSION

In this chapter I have presented data that suggest that grazed phytoplankton dominate under high grazing pressure in summer when nutrients are hardly detectable as measured at macroscales. If one assumes antagonistic impact of herbivory on phytoplankton, this appears paradoxical, as non-grazed phytoplankton would be expected to dominate. To account for this phenomenon I have proposed a mechanism for the positive coexistence of herbivores and their preferred algal prey, which include bacteria, HNF, and viruses taking part in the cycling of nutrients. This mechanism can explain another paradox, namely red tides occurring under quiescent conditions of a stratified watermass and apparently low nutrients, and the autumn bloom of red tide species when nutrients are lost from the euphotic zone. Furthermore, predator-prey synergism in the plankton community may provide a mechanistic explanation for resilience and abrupt and persistent ecosystem shifts in relation to gradually changing environmental condition (eutrophication and temperature) as observed repeatedly along the Norwegian Skagerrak coast. Such shifts can be classified as bifurcations. The proposed mechanism suggests that herbivores exert a strong influence on the algal community. However, in contrast to the traditional perspective of top-down control in ecosystems, the relationship between predator and prey is positive. Hence, this form of predator-prey interaction is potentially a new paradigm in aquatic ecology, and as such, a building block on which to construct new ecological theories. Some potential implications of predator-prey synergism in the plankton community are presented in Chapter 7.

References

Andersen, T., Hessen, D.O., 1991. Carbon, nitrogen, and phosphorous content of freshwater zooplankton. Limnol. Oceanogr. 36, 807–814.

Azam, F., Fenchel, T., Field, J.G., Gray, J.S., Meyer-Reil, L.A., Thingstad, F., 1983. The ecological role of water-column microbes in the sea. Mar. Ecol. Prog. Ser. 10, 257–263.

Banse, K., 1995. Zooplankton: pivotal role in the control of the ocean productivity. ICES J. Mar. Sci. 52, 265–277.

Berg, H.C., Purcell, E.M., 1977. Physics of chemoreception. Biophys. J. 20, 193–219.

Bergelson, J., Crawley, M.J., 1992. Herbivory and *Ipomopsis aggregata*: the disadvantage of being eaten. Am. Nat. 139, 870–882.

Bidle, K.D., Falkowski, P.G., 2004. Cell death in planktonic, photosynthetic microorganisms. Nat. Rev. Microbiol. 2, 643–655.

Blackburn, N., Fenchel, T., Mitchell, J., 1998. Microscale nutrient patches in planktonic habitats shown by chemotactic bacteria. Science 282, 2254–2256.

Bond, W.J., Keeley, J.E., 2005. Fire as a global "herbivore": the ecology and evolution of flammable ecosystems. Trends Ecol. Evol. 20, 387–394.

Bond, W.J., Midgley, G.F., 1995. Kill thy neighbor – an Individualistic argument for the evolution of flammability. Oikos 73, 79–85.

Bongiorni, L., Magagnini, M., Armeni, M., Noble, R., Danovaro, R., 2005. Viral production, decay rates, and life strategies along a trophic gradient in the North Adriatic Sea. Appl. Environ. Microb. 71, 6644–6650.

Broom, D.M., Arnold, G.W., 1986. Selection by grazing sheep of pasture plants at low herbage availability and responses of plants to grazing. Aust. J. Agric. Res. 37, 527–538.

Buskey, E., 1997. Behavioral components of feeding selectivity of the heterotrophic dinoflagellate Protoperidinium pellucidum. Mar. Ecol. Prog. Ser. 153, 77–89.

Calbet, A., Landry, M.R., 2004. Phytoplankton growth, microzooplankton grazing, and carbon cycling in marine systems. Limnol. Oceanogr. 49, 51–57.

Carpenter, S.R., Kitchell, J.F., Hodgson, J.R., 1985. Cascading trophic interactions and lake productivity. Bioscience 35, 634–639.

Chopin, M., Chopin, A., Bidnenko, E., 2005. Phage abortive infection in lactococci: variations on a theme. Curr. Opin. Microbiol. 8, 473–479.

Currie, D.J., 1984. Microscale nutrient patches: do they matter to the phytoplankton? Limnol. Oceanogr. 29, 211–214.

Dahl, E., Johannessen, T., 1998. Temporal and spatial variability of phytoplankton and Chl a—lessons from the south coast of Norway and the Skagerrak. ICES J. Mar. Sci. 55, 680–687.

DeMott, W.R., 1988. Discrimination between algae and artificial particles by freshwater and marine copepods. Limnol. Oceanogr. 33, 397–408.

DeMott, W.R., 1989. Optimal foraging theory as a predictor of chemically mediated food selection by suspension-feeding copepods. Limnol. Oceanogr. 34, 140–154.

Dolan, J.R., McKeon, K., 2005. The reliability of grazing rate estimates from dilution experiments: have we over-estimated rates of organic carbon consumption by microzooplankton? Ocean Sci. 1, 1–7.

Dortch, Q., Clayton, J.R., Thoresen, S.S., Ahmed, S.L., 1984. Species differences in accumulation of nitrogen pools in phytoplankton. Mar. Biol. 81, 237–250.

Doubell, M.J., Yamazaki, H., Li, H., Kokubu, Y., 2009. An advanced laser-based fluorescence microstructure profiler (TurboMAP-L) for measuring bio-physical coupling in aquatic systems. J. Plankton Res. 31, 1441–1452.

Elser, J.J., Urabe, J., 1999. The stoichiometry of consumer-driven nutrient recycling: theory, observations, and consequences. Ecology 80, 745–751.

Fenchel, T., 1988. Marine planktonic food chains. Annu. Rev. Ecol. Syst. 19, 19–38.

Fenchel, T., 2008. The microbial loop-25 years later. J. Exp. Mar. Biol. Ecol. 366, 99–103.

Forde, A., Fitzgerald, G.F., 1999. Bacteriophage defence systems in lactic acid bacteria. Antonie Van Leeuwenhoek 76, 89–113.

Fransz, H.G., Colebrook, J.M., Gamble, J.C., Krause, M., 1991. The zooplankton of the North Sea. Neth. J. Sea Res. 28, 1–52.

Fuhrman, J.A., 1999. Marine viruses and their biogeochemical and ecological effects. Nature 399, 541–548.

Gordon, I.J., Lindsay, W.K., 1990. Could mammalian herbivores "manage" their resources? Oikos 59, 270–280.

Govorunova, E.G., Sineshchekov, O.A., 2005. Chemotaxis in the green flagellate alga *Chlamydomonas*. Biochemestry-Moskow 70, 717–725.

Granéli, E., Carlsson, P., Olsson, P., Sundström, B., Granéli, W., Lindahl, O., 1989. From anoxia to fish poisoning: the last 10 years of phytoplankton blooms in Swedish marine waters. In: Cosper, E.M., Bricelj, V.M., Carpenter, E.J. (Eds.), Novel Phytoplankton Blooms, Causes and Impacts of Recurrent Brown Tides and Other Unusual Blooms. Springer-Verlag, New York, pp. 407–428.

Griffin, A.S., West, S.A., Buckling, A., 2004. Cooperation and competition in pathogenic bacteria. Nature 430, 1024–1027.

Gunderson, L.H., 2000. Ecological resilience—in theory and application. Annu. Rev. Ecol. Syst. 31, 425–439.

Hansen, B., Bjørnsen, P.K., Hansen, P.J., 1994. The size ratio between planktonic predators and their prey. Limnol. Oceanogr. 39, 395–403.

Hansen, H.P., Koroleff, F., 1999. Determination of nutrients. In: Grasshoff, K., Kremling, K., Ehrhardt, M. (Eds.), Methods of Seawater Analysis. Wiley-VCH, Weinheim, pp. 159–228.

Hairston, N.G., Smith, F.E., Slobodkin, L.B., 1960. Community structure, population control, and competition. Am. Nat. 94, 421–425.

Holling, C.S., 1973. Resilience and stability of ecological systems. Annu. Rev. Ecol. Syst. 4, 385–398.

Hutchinson, G., 1961. The paradox of the plankton. Am. Nat. 95, 137–145.

Irigoien, X., Flynn, K., Harris, R., 2005. Phytoplankton blooms: a 'loophole' in microzooplankton grazing impact? J. Plankton Res. 27, 313–321.

Johannessen, T., Dahl, E., 1996a. Declines in oxygen concentrations along the Norwegian Skagerrak coast, 1927–1993: a signal of ecosystem changes due to eutrophication? Limnol. Oceanogr. 41, 766–778.

Johannessen, T., Dahl, E., 1996b. Historical changes in oxygen concentrations along the Norwegian Skagerrak coast: reply to the comment by Gray and Abdullah. Limnol. Oceanogr. 41, 1847–1852.

Johannessen, T., Dahl, E., Falkenhaug, T., Naustvoll, L.J., 2012. Concurrent recruitment failure in gadoids and changes in the plankton community along the Norwegian Skagerrak coast after 2002. ICES J. Mar. Sci. 69, 795–801.

Johannessen, T., Dahl, E., Lindahl, O., 2006. Overgrazing of edible algae as a mechanism behind red tides and harmful algal blooms. Afr. J. Mar. Sci. 28, 337–341.
Jumars, P.A., Penry, D.L., Baross, J.A., Perry, M.J., Frost, B.W., 1989. Closing the microbial loop: dissolved carbon pathway to heterotrophic bacteria from incomplete ingestion, digestion and absorption in animals. Deep-Sea Res. 36, 483–495.
Jürgens, K., Wickham, S.A., Rothhaupt, K.O., 1996. Feeding rates of macro-and microzooplankton on heterotrophic nanoflagellates. Limnol. Oceanogr. 41, 1833–1839.
Kerfoot, W.C., Kirk, K.L., 1989. Degree of taste discrimination among suspension-feeding cladocerans and copepods: implications for detrivory and herbivory. Limnol. Oceanogr. 36, 1107–1123.
Kiørboe, T., 1993. Turbulence, phytoplankton cell size, and the structure of pelagic food webs. Adv. Mar. Biol. 29, 1–72.
Kiørboe, T., Nielsen, T.G., 1994. Regulation of zooplankton biomass and productivity in a temperate, coastal ecosystem. 1. Copepods. Limnol. Oceanogr. 39, 493–507.
Lehman, J.T., 1980. Release and cycling of nutrients between planktonic algae and herbivores. Limnol. Oceanogr. 25, 620–632.
Lehman, J.T., Scavia, D., 1982. Microscale patchiness of nutrients in plankton communities. Science 216, 729–730.
Legendre, L., 1990. The significance of microalgal blooms for fisheries and for the export of particulate organic carbon in oceans. J. Plankton Res. 12, 681–699.
Legendre, L., Rassoulzadegan, F., 1995. Plankton and nutrient dynamics in marine waters. Ophelia 41, 153–172.
Lindahl, O., 1995. Long-term studies of primary phytoplankton productivity in the Gullmar fjord, Sweden. In: Skjoldal, H.R., Hopkins, C., Erikstad, K.E., Leinaas, H.P. (Eds.), Ecology of Fjords and Coastal Waters. Elsevier, Amsterdam, pp. 105–122.
Lindahl, O., Hernroth, L., 1983. Phyto-zooplankton community in coastal waters of western Sweden—an ecosystem off balance? Mar. Ecol. Prog. Ser. 10, 119–126.
Loreau, M., 1995. Consumers as maximizers of matter and energy flow in ecosystems. Am. Nat. 145, 22–42.
Mann, K.H., Lazier, J.R.N., 2006. Dynamics of Marine Ecosystems: Biological-Physical Interactions in the Oceans. Wily-Blackwell, Oxford.
Margalef, R., Estrada, M., Blasco, D., 1979. Functional morphology of organisms involved in red tides, as adapted to decaying turbulence. In: Taylor, D., Seliger, H. (Eds.), Toxic Dinoflagellate Blooms. Elsevier, New York, pp. 89–94.
McCarthy, J.J., Goldman, J.C., 1979. Nitrogenous nutrition of marine phytoplankton in nutrient-depleted waters. Science 203, 670–672.
McNaughton, S.J., 1979. Grazing as an optimization process: grass-ungulate relationships in the Serengeti. Am. Nat. 113, 691–703.
McNaughton, S.J., 1984. Grazing lawns: Animals in herds, plant form, and coevolution. Am. Nat. 124, 863–886.
McNaughton, S.J., Banyikwa, F.F., McNaughton, M.M., 1997. Promotion of the cycling of diet-enhancing nutrients by African grazers. Science 278, 1798–1800.
Middelboe, M., Jørgensen, N.O.G., 2006. Viral lysis of bacteria: an important source of dissolved amino acids and cell wall compounds. J. Mar. Biol. Assoc. UK. 86, 605–612.
Mitchell, J.G., Yamazaki, H., Seuront, L., Wolk, F., Li, H., 2008. Phytoplankton patch patterns: seascape anatomy in a turbulent ocean. J. Mar. Syst. 69, 247–253.
Mitra, A., Flynn, K.J., 2006. Promotion of harmful algal blooms by zooplankton predatory activity. Biol. Lett. 2, 194–197.
Nielsen, T.G., Kiørboe, T., 1994. Regulation of zooplankton biomass and production in a temperate, coastal ecosystem. II. Ciliates. Limnol. Oceanogr. 39, 508–519.
Owen, D.F., Wiegert, R.G., 1981. Mutualism between grass and grazers: an evolutionary hypothesis. Oikos 36, 376–378.

Owen-Smith, N., 1987. Pleistocene extinctions: the pivotal role of magaherbivores. Paleobiol. 13, 351–362.
Paige, K.N., Whitham, T.G., 1987. Overcompensation in response to mammalian herbivory: the advantage of being eaten. Am. Nat. 129, 407–416.
Paine, R.T., 1969. A note on trophic complexity and community stability. Am. Nat. 103, 91–93.
Pernthaler, J., 2005. Predation on prokaryotes in the water column and its ecological implications. Nat. Rev. Microbiol. 3, 537–546.
Prins, H., van der Jeugd, H., 1993. Herbivore population crashes and woodland structure in East Africa. J. Ecol. 81, 305–314.
Riegman, R., Kuipers, R.R., Noordeloos, A.A.M., Witte, H.J., 1993. Size-differential control of phytoplankton and the structure of plankton communities. Neth. J. Sea Res. 31, 255–265.
Roques, K.G., O'connor, T.G., Watkinson, A.R., 2001. Dynamics of shrub encroachment in an African savanna: relative influences of fire, herbivory, rainfall and density dependence. J. Appl. Ecol. 38, 268–280.
Rynearson, T.A., Armbrust, E.V., 2004. Genetic differentiation among populations of the planktonic marine diatom *Ditylum brightewellii* (Bacillariophycea). J. Phycol. 40, 34–43.
Rynearson, T.A., Armbrust, E.V., 2005. Maintenance of clonal diversity during a spring bloom of the centric diatom *Ditylum brightwellii*. Mol. Ecol. 14, 1631–1640.
Ryther, J.H., Dunstan, W.M., 1971. Nitrogen, phosphorus and eutrophication in the coastal marine environment. Science 171, 1008–1013.
Sanders, R.W., Caron, D.A., Berninger, U.-G., 1992. Relationships between bacteria and heterotrophic nanoplankton in marine and fresh waters: an inter-ecosystem comparison. Mar. Ecol. Prog. Ser. 86, 1–14.
Sankaran, M., Hanan, N., Scholes, R., Ratnam, J., Augustine, D., Cade, B., et al., 2005. Determinants of woody cover in African savannas. Nature 438, 846–849.
Scavia, D., Fahnenstiel, G., Davis, J.A., Kreis Jr, R.G., 1984. Small-scale nutrient patchiness: some consequences and a new encounter mechanism. Limnol. Oceanogr. 29, 785–793.
Scheffer, M., Carpenter, S.R., 2003. Catastrophic regime shifts in ecosystems: linking theory to observation. Trends Ecol. Evol. 18, 648–656.
Seymour, J.R., Marcos, Stocker, R., 2009. Resource patch formation and exploitation throughout the marine microbial food web. Am. Nat. 173, E15–E29.
Seymour, J.R., Simó, R., Ahmed, T., Stocker, R., 2010. Chemoattraction to dimethylsulfoniopropionate throughout the marine microbial food web. Science 329, 342–345.
Sinclair, A., Fryxell, J.M., 1985. The Sahel of Africa: ecology of a disaster. Can. J. Zool. 63, 987–994.
Smayda, T.J., 1997a. What is a bloom? A commentary. Limnol. Oceanogr. 42, 1132–1136.
Smayda, T.J., 1997b. Harmful algal blooms: their ecophysiology and general relevance to phytoplankton blooms in the sea. Limnol. Oceanogr. 42, 1137–1153.
Smayda, T.J., 2002. Turbulence, watermass stratification and harmful algal blooms: an alternative view and frontal zones as "pelagic seed banks." Harmful Algae 1, 95–112.
Snyder, J.R., 1984. The role of fire: much ado about nothing? Oikos 43, 404–405.
Sommer, U., 1989. Toward a Darwinian ecology of plankton. In: Sommer, U. (Ed.), Plankton Ecology: Succession in Plankton Communities. Springer-Verlag, Berlin, pp. 1–8.
Sterner, R.W., 1986. Herbivores' direct and indirect effects on algal populations. Science 231, 605–607.
Stocker, R., Seymour, J.R., Samadani, A., Hunt, D.E., Polz, M.F., 2008. Rapid chemotactic response enables marine bacteria to exploit ephemeral microscale nutrient patches. Proc. Natl. Acad. Sci. USA 105, 4209–4214.

Strömberg, C.A.E., 2011. Evolution of grasses and grassland ecosystems. Annu. Rev. Earth Planet. Sci. 39, 517–544.

Suttle, C.A., 2005. Viruses in the sea. Nature 437, 356–361.

Sætre, R. (Ed.), 2007. The Norwegian Coastal Current—Oceanography and Climate. Tapir Academic Press, Trondheim.

Thingstad, T.F., 1998. A theoretical approach to structuring mechanisms in the pelagic food web. Hydrobiologia 363, 59–72.

Thingstad, T.F., 2000. Elements of a theory for the mechanisms controlling abundance, diversity, and biogeochemical role of lytic bacterial viruses in aquatic systems. Limnol. Oceanogr. 45, 1320–1328.

Thingstad, T., Sakshaug, E., 1990. Control of phytoplankton growth in nutrient recycling ecosystems. Theory and terminology. Mar. Ecol. Prog. Ser. 63, 261–272.

Turner, J.T., 2002. Zooplankton fecal pellets, marine snow and sinking phytoplankton blooms. Aquat. Microb. Ecol. 27, 57–102.

Turner, J.T., Tester, P.A., 1997. Toxic marine phytoplankton, zooplankton grazers, and pelagic food webs. Limnol. Oceanogr. 42, 1203–1214.

Verity, P.G., 1991. Feeding in planktonic protozoans: evidence for non-random acquisition of prey. J. Eukaryot. Microbiol. 38, 69–76.

Weiler, C.S., Eppley, R.W., 1979. Temporal pattern of division in the dinoflagellate genus Ceratium and its application to the determination of growth rate. J. Exp. Mar. Biol. Ecol. 39, 1–24.

Weinbauer, M.G., 2004. Ecology of prokaryotic viruses. FEMS Microbiol. Rev. 28, 27–181.

West, S.A., Diggle, S.P., Buckling, A., Gardner, A., Griffin, A.S., 2007. The social lives of microbes. Annu. Rev. Ecol. Evol. Sci. 38, 53–77.

Wilson, D.S., 1973. Food size selection among copepods. Ecology 54, 909–914.

Wynne-Edwards, V.C., 1986. Evolution Through Group Selection. Blackwell Scientific Publications, London.

Yasumoto, T., Underdal, B., Aune, T., Hormazabal, V., Skulberg, O.M., Oshima, Y., 1990. Screening for haemolytic and ichtyotoxic components in *Chrysochromulina polylepis* and *Gyrodinium aureolum* from Norwegian waters. In: Granèli, E., Sundström, B., Edler, L., Anderson, D.M. (Eds.), Toxic Marine Phytoplankton. Elsevier, New York, pp. 436–440.

Young, W.R., Roberts, A.J., Stuhne, G., 2001. Reproductive pair correlations and the clustering of organisms. Nature 412, 328–331.

CHAPTER

7

Ecological Implications of Predator-Prey Synergism in Marine Ecosystems

7.1 INTRODUCTION

In Chapter 6, a conceptual model was proposed for the mutually beneficial relationships among herbivorous zooplankton, their algal prey, and microorganisms facilitating cycling of nutrients, bacteria, heterotrophic nanoflagellates (HNF), and viruses. Such positive relationships between trophic levels were termed predator-prey synergism (hereafter synergism), defined as predator-prey relationships enhancing abundances of both predator and prey. The basic idea underlying synergism in plankton communities is that the organisms that are being grazed (e.g., palatable phytoplankton) gain competitive advantages over non-grazed competitors by sacrificing parts of their clonal populations in order to obtain resources for continuous growth. This is facilitated through cycling of nutrients. The mechanism is based on the perception of the "selfish gene" as clonal copies of phytoplankton, bacteria, and HNF that are left intact, continue to reproduce, and maintain growing populations.

It is recognized that such mutually beneficial coexistence of predator and prey contrasts with the prevailing perception of the predator-prey relationship generally considered antagonistic, that is, that predators have a negative impact on the abundance of their prey. For example, negative interactions between trophic levels are fundamental aspects of "the world is green" hypothesis for terrestrial systems (Hairston et al., 1960) and the cascade hypothesis in aquatic environment (e.g., Carpenter et al., 1985). Synergistic interaction in plankton communities potentially represents a new paradigm in aquatic ecology and as such, a building

From an Antagonistic to a Synergistic Predator Prey Perspective.
DOI: http://dx.doi.org/10.1016/B978-0-12-417016-2.00007-5

block for revising ecological theories. Here, some implications of synergism in plankton are investigated, including annual succession patterns, processes underlying declining resilience rendering marine ecosystems vulnerable to bifurcations (i.e., persistent regime shifts induced by gradually changing environmental conditions), impact of global warming on marine ecosystems, and consequences for the production and harvesting potential of marine resources and thus for fisheries management. In addition, the potential inclusion of higher trophic levels in synergistic interactions is evaluated.

The synergism hypothesis emerged as an alternative predator-prey model in plankton communities to account for phenomena that appeared paradoxical under an antagonistic predator-prey model: dominance of edible phytoplankton under high grazing pressure, red tides occurring in apparently nutrient depleted water, and repeated observation of bifurcations. The synergism hypothesis was developed on the basis of empirical information and has yet to be tested. Extensive literature research did not reveal alternative mechanisms leading to rejection of the synergism hypothesis. Of particular interest are experiments in ponds and shallow lakes showing evidence of food web cascade effects from fish to phytoplankton, following removal or introduction of predators, thus showing negative impact of predation (e.g., Carpenter et al., 1987). Because such observations might provide sufficiently strong evidence to reject the synergism hypothesis, I initially review relevant limnic studies to determine if they actually constitute a sufficient basis for rejection.

7.2 LIMNIC STUDIES

Limnic systems are considered particularly suitable for ecological studies because of their clear boundaries and manageable spatial scales, and that they also display quick responses to experimental manipulation of the plankton community (Sommer, 1989). Limnic and marine ecosystems share many characteristics; hence limnic studies could potentially shed light on the synergism hypothesis primarily developed for marine pelagic ecosystems.

Limnic studies have mainly revolved around the classical question of bottom-up versus top-down control in ecosystems. However, despite many studies, no consensus has been reached about controlling mechanisms (e.g., DeMelo et al., 1992; Carpenter and Kitchell, 1992; Elser et al., 1998). Carpenter (1996, p. 679) argues that the lack of consistency first and foremost illustrates the problem of inappropriate spatial and temporal scales, that "most of the crucial questions of applied ecology are not open to attack by microcosms," and "that field studies at the scale of the environmental problem are essential when phenomena of interest cannot be bottled." Pimm (1991) argues that ecological experiments are almost invariably too limited in terms of the number of species as well as the spatial and temporal scales to be realistic.

It seems reasonable to assume that coevolution of predator and prey is an essential aspect of synergism or indeed any other theory. Hence, data from realistic ecosystems comprising coevolved organisms are probably needed to test the synergism hypothesis. Obviously, microcosms are unlikely to be realistic, but the realism of many full-scale and mesoscale limnic studies is also disputable. A résumé of Carpenter's (2003) historical survey of changes in Lake Mendota, Wisconsin, may be useful to illustrate this point. Lake Mendota has been used in a number of studies of ecological mechanisms in lakes (e.g., Carpenter, 2003; Vanni et al., 1990; Vanni and Temte, 1990). The first written descriptions of Lake Mendota dating back 200 years ago describe a brilliantly clear lake with crystalline blue water. In the 1880s the water had become turbid and green, and newspapers regularly reported choking blooms of algae, foul odors, and fish mortality. Around that time carp was introduced as a game and food fish. Dramatic changes were observed throughout the twentieth century, connected with hydrological regulations, draining of wetlands, introduction of artificial fertilizers in agriculture, changes in municipal sewage management, introduction of alien species, and extermination of indigenous species. During the twentieth century the turnover rates for fish species averaged about 1.8 species added and 2.4 species lost per decade.

Obviously, it is a far cry from the crystal clear water of Lake Mendota 200 years ago to the present highly altered ecosystem as a result both of biological and environmental changes. The coevolutionary aspect is probably long gone, as are the environmental conditions for the indigenous species. It is questionable whether the present ecosystem of Lake Mendota is suitable for studying natural ecosystem processes (interactions between coevolved organisms subjected to natural selection on realistic evolutionary timescales). Although the ecological changes in Lake Mendota may be more extensive than in many other lakes, most lakes in populated areas have been subjected to one or several of the environmental and biological changes described for this lake. In particular, there is a long history of introducing new species in lakes as game and food fishes (Eby et al., 2006). While studies of manipulated lakes are important since "man-made" ecosystems also need to be managed, any inference from such ecosystem studies as to how coevolved organisms interact in natural ecosystems may be deceptive.

Another potential problem with limnic (and marine) experimental studies is that synergism in plankton depends on growth rates in phytoplankton, bacteria, and HNF balance grazing rates. For example, both overgrazing (e.g., by overstocking herbivores) and undergrazing (e.g., understocking herbivores or too much nutrients) of edible phytoplankton may lead to increase in the biomass of little- and non-grazed species, whereas balanced designed experiments will result in low algal biomass and the dominance of grazed species (see Chapter 6 for more detailed mechanistic explanation). Hence, the impact of grazing on the phytoplankton community may be widely different depending on whether herbivores are understocked, overstocked, or balanced relative to the phytoplankton community. For obvious reasons many grazing experiments use overstocking of herbivores; for example, Bergquist et al. (1985, p. 1039) state: "The high zooplankton densities were used to elicit strong, rapid responses from the phytoplankton while minimizing enclosure effects." Biomanipulation experiments, in which top predators are introduced or removed or nutrients are added, may also potentially elicit widely different responses in the phytoplankton community depending on how the balance between herbivores and phytoplankton is affected. Examples of use of very strong doses in biomanipulation experiments include close to total elimination of herbivores by planktivores (e.g., Lynch and Sharpiro, 1981) and nearly total elimination of planktivores by piscivores (e.g., Carpenter et al., 1987). Consequently, both ecological experiments using strong doses and studies of non-coevolved organisms may be highly deceptive and cannot be used to reject the synergism hypothesis.

7.3 THE SEASONAL SUCCESSION IN THE PLANKTON COMMUNITY

This section aims to interpret some aspects of succession in the light of synergism as a basis for evaluating the impact that the next trophic level, the planktivores, has on the pelagic community structure and also how eutrophication may impact plankton and reduce resilience (sensu Holling, 1973) of marine ecosystems. The latter is related to repeated incidents of abrupt and persistent recruitment collapses in gadoid fishes along the Norwegian Skagerrak coast (Chapter 2) that were attributed to abrupt shifts in the plankton community as a result of gradual eutrophication (Chapters 3–5; Johannessen et al., 2012). It should be noted that the seasonal succession described here pertains to spring blooming systems in which both new and recycled nutrients are important for primary production (Chapter 6). In upwelling systems new nutrients are probably more important, whereas in oligotrophic waters recycled nutrients predominate (Cushing, 1989).

A principal outline of the annual pattern in phytoplankton biomass in terms of Chl *a*, primary productivity and nutrients on the south coast of Norway as observed during the period 1990–1999, is presented in Fig. 7.1 (modified from Fig. 6.2). Here, only aspects relevant to the seasonal succession are reviewed (see Chapter 6 for more details). In spring, larger diatoms escape grazing control and bloom due to low over-wintering stocks of mesozooplankton (Riegman et al., 1993). The spring bloom terminates upon depletion of allochthonous nutrients in the euphotic layer. After the spring bloom, grazing control sets in and phytoplankton and herbivores coexist in a synergistic balance where the growth rate in phytoplankton balances grazing. Primary productivity almost doubles from spring to summer. This is possible because nutrients from deeper water are mixed into the euphotic layer, where they are captured and recycled in the plankton community. As sedimentation of phytoplankton is generally low during summer and the algal biomass does not change notably, grazing of algae must increase proportionally to that of primarily productivity (follows from Eq. 6.1, Chapter 6). In agreement with this, the annual pattern in total copepod biomass follows closely that of primary productivity in these waters (Fransz et al., 1991). Also, as primary productivity and grazing rates almost double from spring to summer, so too must the growth rate in phytoplankton. Algal growth rates are generally higher in warmer water, but within the actual temperature range (average 5.9°C in April and 12.2°C in June, daily measurements at 1 m depth 1990–1999) there

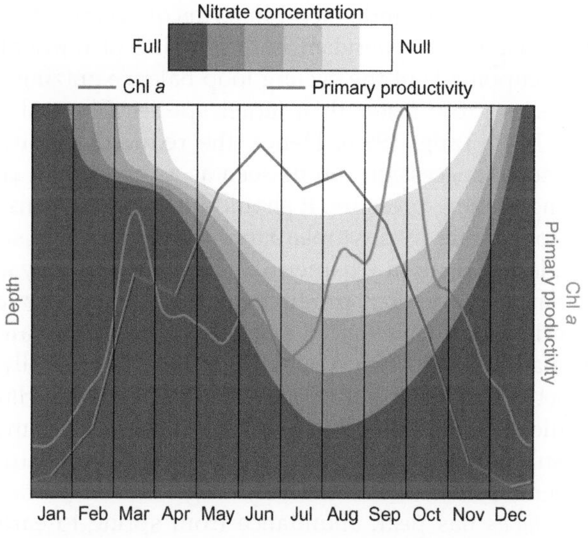

FIGURE 7.1 Principal outline of the patterns in biomass (Chl *a*), primary productivity and nutrient concentration (nitrate) during the seasonal succession. Nutrient concentration relates to depth. *(Modified from Fig. 6.2.)*

are only minor differences (Banse, 1995; Eppley, 1972). Hence, the higher temperature in summer cannot account for the increase in growth rates in phytoplankton.

Unfortunately, flagellates that probably are the main contributors to primary productivity during summer (Dawes, 1998; unpublished counts of total flagellate abundances in these waters by IMR 1989–2001) have not been monitored in detail in these waters. However, there is evidence of decreasing size in algal cells during the seasonal succession in spring blooming systems (Smayda, 1980) and a corresponding reduction in mean size of copepods from spring to summer, both at the community level (Beaugrand et al., 2003) as well as for individual species (Warren et al., 1986, and references therein). This reduced size in algal cells is normally ascribed to small cells being better competitors for nutrients at low concentrations due to their high ratio of surface to volume (Smayda, 1980). However, at microscales, there is evidence of higher concentrations of nutrients from zooplankton exudation and excretion (Blackburn et al., 1998; Lehman and Scavia, 1982; Seymour et al., 2009), and high uptake rates coupled with the ability to store nutrients make it possible for phytoplankton to maintain nearly maximum rates of growth at media concentrations that cannot be quantified with traditional analytical techniques (McCarthy and Goldman, 1979). Accordingly, high primary productivity coupled with low algal biomass in summer implies that algal growth rates must be high and nutrients for algal growth adequate.

Alternatively, the change from larger to smaller algal cells during the seasonal succession can be understood in terms of synergism in plankton. A fundamental aspect of synergism is the cycling of nutrients and that growth in all components of the cycling loop balance grazing. In general, small algal species grow faster than larger species (Kagami and Urabe, 2001; Malone, 1980; Tang, 1995). Hence, the reduced size in algal cells during the seasonal succession can be seen as a response in phytoplankton to increasing grazing pressure. It should be noted that this pertains to the average growth rate in phytoplankton with synergistic strategy (see various strategies in algae in Chapter 6). There are probably a variety of niches with variable primary productivity and from physical (fronts, upwelling, eddies) and biological processes (predation and nutrient release from planktivorous fish schools). As there is generally a positive relationship between predator and prey size in plankton (Hansen et al., 1994), the reduced size in zooplankton from spring to summer corroborates the less-substantiated reduction in phytoplankton size. In agreement with this general picture, the relatively large and important copepod *Calanus finmarchicus* has peak abundance from spring to early summer, whereas the same size copepod, *C. helgolandicus*, has peak abundance in autumn in these waters (Planque and Fromentin, 1996), corresponding to periods with relatively low primary productivity (Fig. 7.1).

There is also another important implication of increased primary productivity without a corresponding increase in the algal biomass. In general, synergistic herbivores may dampen negative impacts of suboptimal environmental conditions for their algal prey (Chapter 6). However, larger synergists are apparently not able to maintain their position when primary productivity increases from spring to summer. Hence, there are two questions that need to be explored in relation to the change in growth rate of phytoplankton rather than algal biomass. First, what prevents small synergists from predominating in spring when primary productivity is low, and second, what prevents larger synergists from maintaining their position under increasing primary productivity?

The answers to these questions can be illustrated by an example. For simplicity, it is assumed that primary productivity doubles from spring to summer, that there are small algae that are half the biomass of large algae, and that the small algae grow twice as fast as the large. Under this scenario, a doubling of the algal biomass of large cells from spring to summer would be sufficient to balance grazing, and conversely, in spring half the biomass of small algae of that in summer would balance grazing. Starting out, immediately after the spring bloom with a plankton community of small synergists and thereby half the biomass of algae when compared with a community consisting of large phytoplankton. Nutrients from deeper water will be mixed into the euphotic layer in pulses, and, as there are high concentrations of nutrients just below the euphotic layer shortly after the spring bloom, early mixing events will supply more nutrients than events following later in the season. If an early mixing event provides more nutrients than the low algal biomass small algae can sequester, nutrients will become available for other algae, including the large synergists. In contrast, a phytoplankton community of larger synergists will have twice as high biomass as the small algae and thereby have a higher capacity to sequester nutrients, which can be stored for subsequent growth (Dortch et al., 1984). As the algal community is generally under grazing control after the spring bloom, this suggests that there is indeed sufficiently high algal biomass of palatable synergistic species to sequester nutrient pulses. This example provides a mechanistic explanation as to why the average size of plankton is relatively large in spring. Another factor that may contribute to a higher proportion of larger synergistic plankton in spring is the competitive advantages larger herbivores may have by being able to efficiently graze the relatively large spring blooming diatoms (e.g., Reigstad et al., 2000).

We can now turn to the second question: Why are larger synergists not able to maintain their dominance as primary productivity continues to increase toward peak level in summer? To answer this question, we will have to analyze it on the basis of the mechanism that underlies synergism between herbivorous zooplankton and their algal prey (Fig. 6.4).

Phytoplankton form micropatches of genetically identical cells due to asexual reproduction. Zooplankton visit patches with high concentrations of preferred algal prey but graze only a proportion of the cells. Algae that are left intact are fertilized by the grazers and continue to grow. Nutrients that are retained by the grazers are compensated by nutrients from deeper water being mixed into the euphotic layer. Phytoplankton do not, however, depend on a continuous supply of new nutrients because they can accumulate intracellular nutrients, which may serve as a long-term reservoir (Dortch et al., 1984). As phytoplankton have to grow at the same rate as they are being grazed for the system to be in balance, the individual algal patch will on average be visited once the cell abundance has recovered.

The problem arises when the abundance of herbivores increases. Higher frequency of visits will prevent algal patches from recovering their abundance, and overgrazing will take place. This will favor more fast-growing algae that are able to keep up with the higher grazing rates. Hence, increasing biomass of herbivores from spring to summer will result in a continuous change toward more fast-growing algae. Overgrazing of slow-growing algae will prevent these algae from fully utilizing recycled nutrients, which will then become available for the smaller algae and thereby be an additional source of nutrients to compensate for nutrients that are retained by the grazers. As the import of new nutrients gradually decreases toward summer, the "overgrazing-source" of nutrients will probably become increasingly important. A simple illustration of the change in the size spectrum in the plankton community during the seasonal succession is presented in Fig. 7.2. These changes, which should not be taken literally, pertain to synergistic plankton. Large plankton predominate during low primary productivity in spring, medium-size plankton predominate between spring and summer,

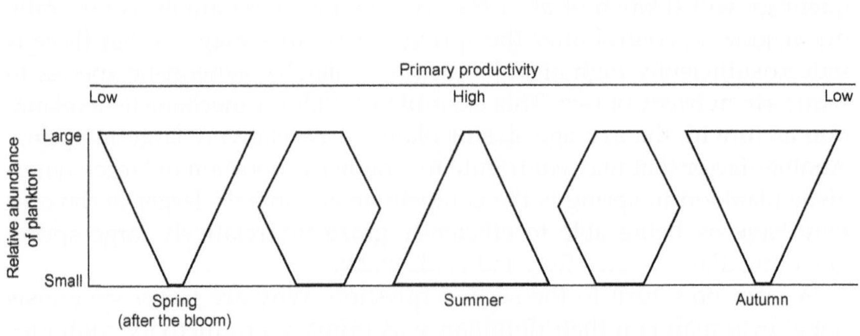

FIGURE 7.2 Principal outline of the size spectrum in synergistic plankton during the seasonal succession as primary productivity changes from low in spring to high in summer and back to low in autumn.

whereas small plankton are the dominating group during peak primary productivity in summer.

In summer, when the import of new nutrients has come to a halt, the opposite process commences. Lack of new nutrients to compensate for nutrients retained by the grazers will result in insufficient nutrient supply and thereby the herbivores overgrazing their algal prey. Severe overgrazing may render recycled nutrients available for little and non-grazed species and cause intensive autumn outbursts (see Section 7.4 for mechanistic explanation). However, the autumn bloom practically vanished after a shift in the plankton community around 2002 (Chapter 5; Johannessen et al., 2012). This shows that the autumn bloom is not inevitable, and underlines the importance of trophic processes for its development (Chapter 6). After the disappearance of the autumn bloom, the seasonal pattern in algal biomass (Chl *a*) has been characterized by a spring bloom followed by relatively stable algal biomass to the end of the productive season in late autumn (Fig. 5.5d), leaving phytoplankton under grazing control also during autumn. Under this scenario the situation during the autumn is similar to that from spring to summer, but reversed. Decreasing primary productivity without a corresponding change in algal biomass implies a shift from small to larger synergists and, obviously, reduced herbivorous biomass and reduced grazing. Indeed, during autumn the total copepod abundance in these waters decreases in parallel with primary productivity (Fransz et al., 1991). This mechanism underlying this shift from small to larger synergist can be explained in terms of overgrazing. Being highly abundant in summer and adapted to high turnover rates, small synergists will be more vulnerable to overgrazing than larger synergists when primary productivity begins decreasing. Overgrazing of small synergistic algae prevents them from sequestering all recycled nutrients, which then become available for other phytoplankton. If at that point the abundance of larger synergistic algae is sufficiently high to sequester the majority of excess nutrients, there will be no autumn bloom of little- and non-grazed dinoflagellates. Overgrazing of progressively larger algal cells will continue to the end of the productive season, accompanied by gradually increasing size in synergistic plankton. It seems reasonable to assume that there are herbivores that have taken advantage of this niche by spawning in late summer so that the offspring will grow in parallel with the benefits of increasing size in synergistic plankton. One such candidate in these waters is *C. helgolandicus*.

In summary, diatoms bloom in spring due to low over-wintering stocks of mesozooplankton grazers. After the spring bloom, herbivorous zooplankton coexist in synergistic balance with their algal prey. Under relatively low primary productivity shortly after the spring bloom, larger and more slow-growing synergists will have a competitive advantage over small and fast-growing synergists because they are able to sequester more of the nutrients that are mixed into the surface layer from high concentration just below the euphotic zone. After the spring bloom, the seasonal succession can be regarded as a continuous process of herbivores overgrazing their algal prey as primary productivity increases toward summer and decreases again in the autumn. Hence, from a synergistic perspective the level of primary productivity is an essential structuring mechanism in marine spring blooming ecosystems: low primary productivity favors large and slow-growing phytoplankton and large herbivorous zooplankton, whereas high primary productivity favors small and fast-growing phytoplankton and thereby small zooplankton. This relationship can be used to evaluate how eutrophication and reduced abundances of planktivores from fishing may affect the plankton community.

7.4 IMPACT OF EUTROPHICATION ON SYNERGISTIC PLANKTON

The seasonal succession in the plankton community can be regarded as a "natural eutrophication" process in which nutrients from deeper water are mixed into the euphotic layer, where they are captured and recycled. This results in increased primary productivity, a corresponding increase in herbivorous biomass and grazing, no increase in algal biomass, but a shift toward more fast-growing, smaller algal cells and smaller herbivores.

The spring bloom of diatoms terminates with the depletion of silicate in the euphotic zone (Chapter 6). As silicate does not change with anthropogenic eutrophication (Gilpin et al., 2004; Jickells, 1998), the magnitude of the spring bloom will probably not be affected. In agreement with this, the algal biomass in spring did not change in relation to decreasing nutrient loads in the coastal waters of Skagerrak (Chapter 5), and in the southern North Sea Wiltshire et al. (2008) found that the spring bloom dynamics have hardly changed at all in spite of considerable environmental changes.

Nitrogen, which is generally the limiting nutrient in these waters (Chapter 6), increased substantially from the 1960s (North Sea Task Force, 1993), reached peak levels in the mid-1990s (Aure et al., 1998, 2010), and has since shown a decreasing trend yet remains elevated when compared with pristine conditions (Frigstad et al., 2013). The increase in primary productivity followed the same pattern as nitrogen (Lindahl et al., 2009). Most of the anthropogenic nutrients along the south coast of Norway originate from the German Bight in the southern North Sea, and this is particularly evident in winter-spring (Aure et al., 1998; Frigstad et al., 2013).

Anthropogenic nutrient loads will result in a faster increase in primary productivity after the spring bloom and an elevated level throughout the productive season when compared with pristine conditions. This, in turn, will speed up the seasonal succession and result in a shorter temporal window for larger and more slow-growing phytoplankton and their grazers.

In summer, elevated primary productivity will be favorable for small and fast-growing algae and smaller grazers. It should be noted though, that this pertains to synergistic phytoplankton. There are, for example, algae with a non-palatable strategy that may bloom in summer (Fig. 8.1), some of which are both large and slow-growing, such as *Ceratium* (Weiler and Eppley, 1979). Such blooms are dead-ends for energy flow into the pelagic food web. Studies of the overall structure of phytoplankton communities, such as size-fractionated Chl *a* without taking into account whether the algae are grazed or not, may be highly deceptive with respect to energy flow in marine ecosystems and in worst case mainly reflect patterns in weed rather than crop.

There are two conditions that may cause little- and non-grazed phytoplankton to bloom in summer and autumn: (1) collapse of synergism due to synergistic herbivores severely overgrazing their preferred algal prey, and (2) excess of allochthonous nutrients (Chapter 6). There are numerous examples from estuaries and lakes, where under highly eutrophic conditions water may stay turbid during most of the productive season due to high concentrations of phytoplankton (e.g., Braarud, 1945; Carpenter, 2003). Such high algal concentrations can be analyzed in light of synergism. The seasonal succession in the plankton community from spring to summer is driven by a continuous process by herbivores overgrazing larger algae that are not able to keep up with the increasing grazing rates. Eutrophication will contribute to higher primary productivity and a shift toward smaller and more fast-growing synergists. There is, however, a limit to algal growth rates (which may vary between different ecosystem states). When this limit is reached, the biomass of the smallest synergists may continue to increase by outcompeting larger synergists in niches that would be present under optimal eutrophic condition. This change in the size-composition in synergistic plankton from an optimal-eutrophic state (probably system dependent) to hypereutrophic state is illustrated in Fig. 7.3. At some point the abundance of synergistic herbivores will become saturated—when the density of herbivores is so high that micropatches of their preferred algal prey are just able to recover before being grazed again. Above the saturation level for synergists, allochthonous nutrients will become available for non-palatable algae, resulting in high algal biomass and discoloration of the water as typically observed under hypereutrophic conditions. This may provide a mechanistic explanation for

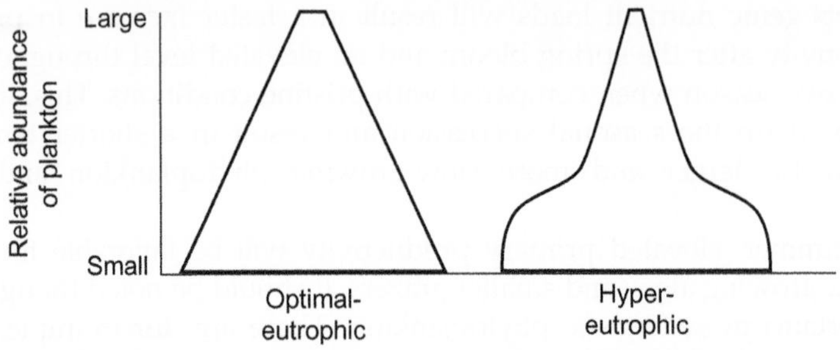

FIGURE 7.3 Principal outline of the size spectrum in synergistic plankton in summer in situations with optimal-eutrophic and hypereutrophic conditions during maximal primary productivity. Productivity changes from low in spring to high in summer and back to low in autumn.

the apparent increase in harmful algal blooms with increasing eutrophication (Anderson et al., 2002; Hallegraff, 1993; Smayda, 1990).

According to this mechanism there will be a clear-water phase after the spring bloom in highly eutrophic waters, which will last until the density of synergists has become saturated. The length of the clear-water phase will be shorter in highly eutrophic water. Interestingly, in a seminal paper about the seasonal succession in the plankton community in lakes, Sommer et al. (1986) describe exactly this pattern for eutrophic lakes, a clear-water phase followed by high algal biomass of non-palatable phytoplankton, among other large and slow-growing dinoflagellates.

The intensive autumn bloom of little- and non-grazed dinoflagellates, *K. mikimotoi* and *Ceratium* spp. (Fig. 7.1), started to appear in Skagerrak in the 1970s (Lindahl and Hernroth, 1983) but practically vanished after 2002, concurrent with other evidence of an abrupt shift in the plankton community (Chapter 5; Frigstad et al., 2013; Johannessen et al., 2012). When such shifts occur, it is postulated that the new community will consist of organisms for which the altered environmental conditions are closest to optimal (Chapter 5). Hence, the absence of the autumn bloom after the shift in 2002 suggests that when environmental conditions are close to optimal, algal blooms of little- and non-grazed species are less common. A mechanistic explanation for autumn blooms of non-grazed dinoflagellates in relation to cultural eutrophication can be deduced from Fig. 7.3. In a system for which the eutrophic state is optimal, there will be a higher proportion of larger synergists than in the system that is hypereutrophic. Therefore, when primary productivity begins decreasing in the late summer and small synergistic algae become overgrazed, the abundance of larger synergistic algae will normally be sufficiently high to sequester excess of recycled nutrients from the overgrazing of small algae. In the over-eutrophic state, on the other hand, there may not be sufficient biomass of larger synergistic algae to sequester excess nutrients, which may therefore become available for the little- and non-grazed dinoflagellates and induce blooms of these algae.

In summary, eutrophication will result in elevated primary productivity throughout the seasonal succession, except during the spring bloom. The temporal window for larger herbivores and their algal prey will be reduced and the annual abundance of larger zooplankton will become lower. The abundance of smaller synergists will increase. Hence, eutrophication will result in a shift toward smaller plankton. Furthermore, evidence suggests that eutrophication will result in increased blooms of little- and non-grazed phytoplankton from synergistic herbivores overgrazing their algal prey, and under more severe eutrophic conditions from "saturation" of synergists rendering excess nutrients available for algal species with a non-palatable strategy.

7.5 IMPACT OF PLANKTIVORES ON SYNERGISTIC PLANKTON

When considered in isolation, the impact that planktivores have on the abundance of their zooplankton prey is obviously negative. However, from an ecosystem perspective it can be argued that this relationship can be turned upside down. It is not only a question of how planktivores affect the abundance of their prey directly, but also how predation will affect the relationship between herbivorous zooplankton and their algal prey. In Section 7.3 it was suggested that the level of primary productivity is an essential structuring mechanism in marine spring blooming ecosystem. The structuring impact of primary productivity is the basis of evaluating the impact planktivores will have on the plankton community.

Planktivores will contribute to reduce primary productivity by different processes. First, planktivores incorporate herbivorous biomass into somatic growth. Second, the relatively large fecal pellets from planktivores sink rapidly out of the euphotic layer. For example, sinking rates $>1000\ \mathrm{m\ d^{-1}}$ of fecal pellets of fish are common (Saba and Steinberg, 2012; Turner, 2002). In addition, the release of dissolved organic compounds from fish fecal pellets is low and does not represent a significant output during the short time required to sink through the water column (Robison and Bailey, 1981). Third, vertical migration in fish and invertebrates also contributes to the removal of carbon and nitrogen from the euphotic zone (Ducklow et al., 2001). In a study of 20 freshwater lakes, Hudson et al. (1999) confirmed that the nutrient supply for lake plankton comes primarily from within the plankton community rather than from external loading or from larger organisms such as fish.

Interestingly, fecal pellets of herbivores grazing diatoms sink faster than when grazing other algae, such as flagellates (Turner, 2002). Although this may be purely random, the consequence is that diatom grazing (diatoms are not synergists) in spring result in a greater loss of nutrients than during flagellate grazing in summer.

All factors that contribute to export of nutrients out of the euphotic zone will reduce the temporal increase in primary productivity from spring to summer and reduce the primary productivity level throughout the productive season. Planktivores will thus counteract the "natural eutrophication" process occurring during the seasonal succession. The total biomass of herbivorous zooplankton will decrease in systems with high levels of planktivores, but as there will be a longer temporal window for larger and more slow-growing synergists, there will be an increase of larger herbivores. The losers will be smaller planktonic synergists and planktivores feeding on these smaller herbivores. An optimal strategy for this group of planktivores would be to reduce the export of nutrients by producing fecal pellets with low sinking rates and not to conduct vertical migration below the euphotic layer (obviously there are also other concerns, such as predation).

The relatively large copepod *C. finmarchicus* constitutes the major biomass component of the mesozooplankton over wide areas of the North Atlantic and is a key prey species for many fishes (Sundby, 2000). For example, late stages of *Calanus* are consumed efficiently by small (≥3–4 cm) 0-group Atlantic cod (*Gadus morhua*) (Chapter 4). They are important prey for adult herring (*Clupea harengus*), mackerel (*Scomber scombrus*), and lesser sandeel (*Ammodytes marinus*) (Prokopchuk and Sentyabov, 2006; Reay, 1970) and staple food for the North Atlantic right whale (*Eubalaena glacialis*) (Baumgartner et al., 2003). High biomass of these planktivores will contribute to reduced primary productivity. The success of *C. finmarchicus* in the North Atlantic may therefore be a result of an optimal primary productivity window opened up by their predators at the expense of smaller herbivores. Small copepods (less than 1 mm) do not provide sufficient energy for small Atlantic cod (mean length 6.1 cm; Fig. 4.5f, Chapter 4). Therefore, it appears unlikely that small copepods are adequate food for larger planktivores such as adult herring and mackerel. This suggests that there is a positive relationship between planktivores and their main prey, larger herbivores, and that synergism comprises three trophic levels in marine ecosystems: phytoplankton (including bacteria, HNF, and viruses), herbivorous zooplankton, and planktivores. The positive relationship has substantial implications for management of marine resources (see Chapter 8).

The evolution of a positive relationship between large herbivores and planktivores does not depend on group selection. The reproductive potential (fecundity) of practically all organisms is much higher than what is attainable (Darwin, 1859), particular in marine organisms. For example, Marchall and Orr (1952) found that female *C. finmarchicus* produced up to 586 eggs during one season. Therefore, even though predator avoidance may be highly developed in copepods, predation is inevitable. Since larger herbivores will benefit from balanced predation due to a reduced level of primary productivity, it is possible that the selection pressure for high fecundity may be a result of both intraspecific and interspecific competition, that is, a high surplus of small offspring is an optimal strategy in competition with conspecifics, and, as predation on the numerous offspring will inevitably be high, genes coding for high fecundity in larger zooplankton may be beneficial in the competition with smaller and more fast-growing plankton.

Having included planktivores as components of the synergistic interaction in the pelagic food web, it could be tempting to consider predation on planktivores as a synergistic strategy too, such as by preventing planktivores from overexploiting their resources. However, there is another and intuitively more appealing mechanism to prevent such overexploitation. At very high concentrations planktivores may include a higher proportion of planktivorous offspring in their diet and thereby reduce recruitment and a further increase in the planktivorous biomass. This is in agreement with the Ricker stock-recruitment curve (Ricker, 1954). Below the potential level by which the planktivores overexploit their herbivorous resources, there will be a direct negative impact of predation on the planktivores, as there are no obvious compensatory mechanisms for the larger herbivores. On the contrary, reduced stocks of planktivores will result in an increase in the total biomass of herbivores and a corresponding increase in primary productivity, which in turn will favor smaller plankton. Hence, predation on planktivores is probably an alternative strategy to synergism.

And indeed, synergism is not the only viable strategy. In phytoplankton, for example, there are non-synergistic diatoms that outgrow grazers and bloom in nutrient-rich and turbulent water and non-palatable opportunistic species that bloom in situations where synergistic herbivores severely overgraze their preferred algal prey or bloom in excess of allochthonous nutrients. There are probably alternative strategies at the other trophic levels as well. Because there are alternative stable states in marine ecosystems and synergism is the mechanism that makes the various states resilient (Chapter 6), such alternative strategies appear not to compromise the integrity of the systems and can therefore be considered as loopholes in a synergistic realm.

In summary, it is suggested that synergism is important at the three lowest trophic levels of the pelagic food web: phytoplankton and bacteria, HNH, and viruses taking part in cycling of nutrients; herbivorous zooplankton; and planktivores.

7.6 IMPACT OF FISHING ON SYNERGISTIC PLANKTON AND ENERGY FLOW

Reduced biomass of planktivores from fishing will result in increased primary productivity and thereby a shift from larger to smaller herbivores. As small herbivores are unlikely to provide sufficient energy for larger planktivores, reduced size in herbivores probably results in a higher proportion of carnivorous prey in their diet (both invertebrates and small fish). The theoretical impact of such a change can be analyzed by Eq. 7.1.

$$B_h = kB_H = \langle fB_H(100/E)\rangle + ((1-f)B_H) \tag{7.1}$$

where B_h is the biomass of herbivores needed to provide planktivores with sufficient energy, B_H is the biomass consumed by the planktivores when 100% of the diet consists of herbivores, k is the relative change in herbivore biomass with a proportion of carnivores in the diet, f is the proportion of carnivores in the diet, and E is the energy transfer efficiency between trophic levels. E is generally considered to be 10% (e.g., Pauly and Christensen, 1995), but this figure can be up to 30% in carnivorous zooplankton (Greenstreet et al., 1997; and references therein). By setting $B_H = 1$ the relative increase in herbivorous biomass to satisfy the planktivores in relation to an increasing proportion of carnivore in their diet can be estimated. With $E = 10\%$ and $f = 0.25$, 0.5, 0.75, and 1.0, the herbivorous biomass would have to increase by a factor (k) of 3.25, 5.5, 7.75, and 10, respectively, relative to a diet of exclusively herbivores. With $E = 30\%$ and the same f, the increase would be 1.58, 2.17, 2.75, and 3.33. In a similar way it can be estimated how much a constant herbivorous biomass (set to 1) can support planktivores with an increasing proportion of carnivorous diet. With $E = 10\%–30\%$ and $f = 0.25$, 0.50, 0.75, 1.00, the supported biomass would be 0.775–0.825, 0.55–0.65, 0.325–0.475, respectively.

There is also another aspect with the change from herbivorous to a carnivorous diet. *C. finmarchicus* and other (mainly) herbivorous copepods accumulate wax esters (lipids) from phytoplankton, which they store in oil sacks as energy reserves (Lee et al., 2006). Lipids may comprise more than 50% of the dry weight of *C. finmarchicus*, whereas in carnivorous copepods the lipid content is much lower (Sargent and Falk-Petersen, 1988). The low energy content of carnivorous prey was clearly demonstrated by the fact that relatively larger hyperbenthic prey such as fish and shrimps did not provide sufficient energy for small young-of-the-year cod, whereas the condition and survival of small cod increased significantly with higher proportions of pelagic prey in their diet, in particular *Calanus* (greater than 2 mm) but also slightly smaller copepods (1–2 mm; Chapter 4). Hence, a stomach full of *Calanus* will contain much more energy than a stomach with carnivorous prey.

There is an interesting story from the North Sea that may shed light on the impact planktivores may have on the plankton community. In the mid-1960s there were huge stocks of pelagic fish: the spawning stock of herring was estimated to approximately 2.5 million tons (ICES, 2012a) and mackerel to greater than 3 million tons (Hamre, 1980; ICES, 2012b). At that time *C. finmarchicus* was highly abundant and was totally dominating among larger copepods (Reid et al., 2003). Heath (2007) considered mackerel to be a pelagic piscivore. However, mackerel feed opportunistically, and in the Norwegian Sea *C. finmarchicus* totally dominated the diet (Prokopchuk and Sentyabov, 2006). In the latter half to the 1960s both herring and mackerel collapsed from overfishing. Having been relieved of the heavy predation pressure from these huge pelagic stocks, one would, in an antagonistic predator-prey perspective, expect the abundance of *C. finmarchicus* to increase, even in a situation with unfavorable physical environmental condition for the species. In contrast to this expectation, the abundance of *C. finmarchicus* was drastically reduced (Reid et al., 2003) in parallel with the collapse of herring and mackerel. There is an important lesson to be learned from this—abundant stocks of planktivorous fishes do not pose a threat to their main prey. The observation from the North Sea is in agreement with synergism interactions between planktivores and large herbivores, namely that the abundance of their larger zooplankton prey will increase in parallel with the abundances of their predators.

Heath (2007) estimated that the consumption of zooplankton production by fish and other planktivorous predators dropped from 25 gC m^{-2} $year^{-1}$ in the early 1970s to 19 g C m^{-2} $year^{-1}$ in the 1990s. Heath's starting year, 1973, was after the collapse of both herring and mackerel. Hence, it seems reasonable to assume that the reduction in zooplankton consumption would have been substantially higher if the mid-1960s had been included. According to the synergism hypothesis, this should result in a shift from larger to smaller herbivorous zooplankton. Indeed, this is what has been observed in the North Sea (see Beaugrand et al, 2003). One result, though, may appear to be in conflict with the synergism hypothesis. The total biomass of calanoid copepods decreased in the 1990s (Beaugrand et al., 2003), whereas the synergism hypothesis predicts that the herbivorous biomass should increase because of reduced predation. However, a very important group of herbivores has not been systematically monitored in the North Sea or in general (Turner, 2004), namely microzooplankton (less than 0.2 mm), which are the main consumers of primary production across a wide range of marine ecosystems (Calbet and Landry, 2004). An increase in these small herbivores would be consistent with the synergism hypothesis.

7.7 EUTROPHICATION, TEMPERATURE INCREASE, OVERFISHING, RESILIENCE, AND BIFURCATIONS

Locally along the Norwegian Skagerrak coast repeated events of abrupt and persistent recruitment collapses in gadoid fishes were observed in relation to increasing eutrophication (Chapter 2) and increasing temperature (Chapter 5). Comprehensive testing in the field using Atlantic cod (*Gadus morhua*) as a model species provided substantial evidence to suggest that the collapses were caused by abrupt changers in the plankton community (Chapter 4). Direct (copepods and phytoplankton) and indirect (oxygen concentrations) evidence supported this conclusion (Chapter 5; Johannessen et al., 2012). Eutrophication is generally a gradual process (Chapter 2). Hence, abrupt changes in the plankton community as a result of gradual changes imply that the plankton community may not respond in a gradual dose-response manner but may shift abruptly from one stable state to another, as in bifurcation (Chapter 1). In one instance nutrient loads were reduced well below the level at which the shift occurred (Chapter 2), without the system returning to the preceding state. Hence, marine plankton communities appear to be highly resilient. Synergism between herbivorous zooplankton and their algal prey was suggested to be the mechanism that causes marine ecosystems to be resilient (Chapter 6). Because nutrient concentration in itself is not problematic for marine organisms and low oxygen was not the cause of the bifurcations, it was suggested that altered competition in plankton was the cause of the shifts (Chapter 5). During optimal environmental conditions for a system, the competitive advantages for the organisms of the system are high and resilience accordingly high. In a situation where the environmental conditions change in favor of other organisms, the resilience of the system is reduced and the system will become vulnerable to bifurcations from physical or biological perturbations (e.g., mass mortality from extreme temperatures or toxic algal blooms) that the system under optimal conditions could withstand (Fig. 1.1).

According to the synergism hypothesis, increased nutrient loads affect competition in plankton by favoring small and fast-growing algae and their grazers due to increased primary productivity and turnover rates. Hence, the synergism hypothesis predicts that eutrophication-induced bifurcations result in a shift to smaller plankton. Unfortunately, no long-term studies of the plankton community were conducted in any of the areas with recruitment collapses, and in short-term studies microzooplankton have been neglected. In a one-year study in the Inner Oslofjord (Fig. 2.1) where the recruitment in gadoids collapses around 1930, Wiborg (1940) found that the copepod community was dominated by small cyclopoid copepods, *Oithona* spp. and *Oncaea borealis* (females 0.5–0.65 mm). These copepods are mainly carnivorous (Castellani et al., 2005; Nielsen and

Sabatini, 1996; Kattner et al., 2003) and thus indicative of a high proportion of primary productivity being consumed by microzooplankton. In a two-year study in the Grenlandfjords (Fig. 4.1) where the recruitment of gadoids collapsed in the mid-1960s (Fig. 2.7), the diet of young-of-the-year cod was dominated by small copepods (less than 1 mm; Fig. 4.5f), which resulted in low survival and poor recruitment. Hence, both of these short-term studies point toward shifts from larger to smaller herbivores and possibly the dominance of microzooplankton. Interestingly though, in both areas exceptionally good recruitment in cod followed advection of *Calanus* into the fjords (Chapters 2 and 4). In conclusion, synergism in the plankton community provides a mechanistic explanation for resilience as well as how eutrophication affects competition in the plankton community (increased primary productivity and turnover rates) and thereby reduces the resilience of the system.

In relation to the concurrent shift in plankton and recruitment failure in gadoid fishes in the coastal waters of Skagerrak in the early 2000s, the total copepod biomass decreased substantially (Chapter 5), whereas primary productivity was relatively unaffected (Lindahl et al., 2009). Because grazing of phytoplankton and cycling of nutrients are important factors for primary productivity (Eq. 6.1), unaltered primary productivity suggests that grazing rates have been approximately the same before and after the shift in the plankton community. Hence the reduced grazing by copepods must have been compensated. As the monitoring conducted by Institute of Marine Research of mesozooplankton (less than 0.2 mm) has not identified increases in the biomass of other plankton groups, an increase in microzooplankton grazing seems to be a likely explanation. The ecosystem shift along the Skagerrak coast in the early 2000s occurred during increasing temperature but decreasing nutrient loads (Chapter 5). Accordingly, different responses were suggested in the ecosystem when compared to the eutrophication-induced shifts. On the other hand, when the shift occurred, the nutrient concentrations were still well above pristine conditions (Frigstad et al., 2013). Hence, both temperature increase and elevated nutrient loads may have contributed to reduce the resilience of the system by favoring phytoplankton adapted to the combination of higher temperature and higher turnover rates, possibly favoring small micozooplankton grazers.

According to the synergism hypothesis, both eutrophication and overfishing of planktivores will result in increased primary productivity. Hence, the ecological impact of these bottom-up and top-down processes are the same. In ecosystems subjected to both eutrophication and overfishing of planktivores, resilience will be reduced by the sum of these two factors. Negative impact of eutrophication can thus be mitigated by management of the fisheries aiming at high stocks of planktivores.

In support of this, there is massive evidence to suggest that pelagic fish stocks were highly abundant prior to the extremely high exploitation of these resources during the latter half of the twentieth century (e.g., Sætersdal, 1980).

7.8 MITIGATING NEGATIVE IMPACTS OF GLOBAL WARMING

With the prospect of global warming, a highly topical question is how marine ecosystem will respond to increasing temperature and reduced pH levels from increasing CO_2 emissions. For organisms that might reach tolerance limits, the answer for the individual species is quite simple. Also, if there had been a gradual dose-response relationship in marine ecosystems, one could have monitored changes and acted to reverse trends before reaching critical stages. However, the long series from the Norwegian Skagerrak coast show that marine ecosystems may shift abruptly between alternative stable states as a result of gradual changes in environmental conditions (Chapters 2 and 5). To predict the consequences of global warming and induce measures to mitigate negative impacts on marine ecosystems, there is a need for a thorough understanding about how marine organisms interact. In Chapter 6 it was argued that the classical perception of antagonistic interactions among planktonic organisms will result in a gradual dose-response relationship and therefore be incompatible with high resilience and abrupt ecosystem shifts. In contrast, the synergism hypothesis provides a mechanistic explanation for resilience and bifurcation.

The synergism hypothesis predicts that global warming may cause abrupt and persistent shifts in the plankton community by the combined impact of increasing temperature and reduced pH affecting competition and thereby reducing the resilience of the system. The vast majority of marine teleost fishes (Leis, 2007) and a high proportion of benthic invertebrates (Pechenik, 1999; Thorson, 1950) depend on planktonic prey during early life stages, and abrupt shifts in the plankton community will therefore probably have substantial impacts on the entire ecosystem. Recruitment collapses, as observed in the gadoids along the Norwegian Skagerrak coast (Chapters 2 and 5), are likely to occur and result in the collapse of indigenous fish stocks. Successful recruitment depends of a number of factors that are a result of evolutionary adaptations, such as spawning location, spawning time, drift routes of larvae, or retention of larvae, behavioral adaptations of larvae and juveniles to reduce predation risk, and so forth. The life cycle of the European eel (*Anguilla anguilla*) illustrates how complicated this adaptation process can be for successful recruitment: Eels spawn in the Sargasso Sea south of Bermuda, the eel larvae drift for several years before entering rivers and brooks to grow up in freshwater systems all along the coasts from the White Sea in northern Russia to northern Africa, and then navigate back up 8000 km to the Sargasso Sea to spawn again (Tesch, 2003). Therefore, it appears unlikely that new species will be able to reproduce efficiently shortly after a bifurcation. In agreement with this, the gadoids have not been replaced by other fishes after the recruitment collapses along the Norwegian Skagerrak coast, in spite of this having occurred approximately 40, 50, and 85 years ago (Chapter 2).

As a consequence, shifts in the plankton community resulting from global warming are likely to result in the collapse of indigenous species without these being efficiently replaced by immigrants. This will result in reduced predation on herbivorous zooplankton, increased primary productivity, a shift toward smaller plankton, a longer food chain, and thereby less-efficient energy transfer to high trophic levels. In addition, there is evidence to suggest that increasing temperature will favor smaller plankton (Atkinson et al., 2003; Daufresne et al., 2009). As evidenced from the Norwegian Skagerrak coast, increased primary productivity may in itself elicit bifurcations. Hence, collapse of fish stocks following a bifurcation caused by global warming will exacerbate the negative impacts and contribute to higher resilience in the new community structure, possibly rendering the system irreversible. The consequences for fish production may thus become very negative.

The observation that the plankton community is highly resilient in spite of high variability in the community structure suggests that within a given state the diversity in the plankton community is sufficiently high to fill available niches. Hence, it seems unlikely that there are keystone species that the plankton community hinges on. Establishing tolerance limits for single species with respect to temperature or pH will therefore probably not be helpful in order to predict bifurcations. Reduced resilience may cause an increase in summer and autumn blooms of non-grazed phytoplankton species (Section 7.4). However, the mechanistic explanation for this is related to increased primary productivity (Fig. 7.3). There are no obvious reasons why blooms of non-grazed species should increase in relation to increasing temperature. Consequently, there may be no warning signs prior to bifurcations caused by global warming. Management of marine ecosystems should therefore aim at maintaining high resilience. This implies identifying factors that may cause competitive changes in plankton and to avoid changing these factors from pristine conditions. In addition to temperature and pH, there are two other obvious factors that affect competition in plankton, namely eutrophication and overfishing of planktivorous fishes, which both cause primary productivity to increase. In order to mitigate negative impacts of global warming, eutrophication should be reduced and planktivorous fish stocks should be kept at high levels, closer to the abundances during pristine conditions.

References

Anderson, D.M., Glibert, P.M., Burkholder, J.M., 2002. Harmful algal blooms and eutrophication: nutrient sources, composition, and consequences. Estuaries 25, 704–726.

Atkinson, D., Ciotti, B.J., Montagnes, D.J.S., 2003. Protists decrease in size linearly with temperature: ca. 2.5%C°. Proc. R. Soc. London B 270, 2605–2611.

Aure, J., Danielssen, D., Svendsen, E., 1998. The origin of Skagerrak coastal water off Arendal in relation to variations in nutrient concentrations. ICES J. Mar. Sci. 55, 610–619.

Aure, J., Danielssen, D., Magnusson, J., 2010. Langtransporterte tilførsler av næringssalter til Ytre Oslofjord 1996–2006. Fisken og havet 4/2010, 1–24 (in Norwegian).

Banse, K., 1995. Zooplankton: pivotal role in the control of ocean production. ICES J. Mar. Sci. 52, 265–277.

Baumgartner, M.F., Cole, T.V., Campbell, R.G., Teegarden, G.J., Durbin, E.G., 2003. Associations between North Atlantic right whales and their prey, *Calanus finmarchicus*, over diel and tidal time scales. Mar. Ecol. Prog. Ser. 264, 155–166.

Beaugrand, G., Brander, K.M., Lindley, J.A., Souissi, S., Reid, P.C., 2003. Plankton effect on cod recruitment in the North Sea. Nature 426, 661–664.

Bergquist, A.M., Carpenter, S.R., Latino, J.C., 1985. Shifts in phytoplankton size structure and community composition during grazing by contrasting zooplankton assemblages. Limnol. Oceanogr. 30, 1037–1045.

Blackburn, N., Fenchel, T., Mitchell, J., 1998. Microscale nutrient patches in planktonic habitats shown by chemotactic bacteria. Science 282, 2254–2256.

Braarud, T., 1945. A phytoplankton survey of the polluted waters of inner Oslo Fjord. Hvalråd. Skr. Sci. Results Mar. Biol. Res. 28, 1–142.

Calbet, A., Landry, M., 2004. Phytoplankton growth, microzooplankton grazing, and carbon cycling in marine systems. Limnol. Oceanogr. 49, 51–57.

Carpenter, S.R., 1996. Microcosm experiments have limited relevance for community and ecosystem ecology. Ecology 77, 677–680.

Carpenter, S.R., 2003. Regime shifts in lake ecosystems: pattern and variation. In: Kinne, O. (Ed.), Excellence in Ecology International Ecology Institute, Oldendorf/Luhe, Book 15.

Carpenter, S.R., Kitchell, J.F., 1992. Trophic cascade and biomanipulation: interface of research and management—a reply to the comment by DeMelo et al. Limnol. Oceanogr. 37, 208–213.

Carpenter, S.R., Kitchell, J.F., Hodgson, J.R., 1985. Cascading trophic interactions and lake productivity. Bioscience 35, 634–639.

Carpenter, S.R., Kitchell, J.F., Hodgson, J.R., Cochran, P.A., Elser, J.J., Elser, M.M., et al., 1987. Regulation of lake primary productivity by food web structure. Ecology 68, 1863–1876.

Castellani, C., Irigoien, X., Harris, R.P., Lampitt, R.S., 2005. Feeding and egg production of Oithona similis in the North Atlantic. Mar. Ecol. Prog. Ser. 288, 173–182.

Cushing, D.H., 1989. A difference in structure between ecosystems in strongly stratified waters and in those that are only weakly stratified. J. Plankton Res. 11, 1–13.

Darwin, C., 1859. On the Origin of Species. Murray, London.

Daufresne, M., Lengfellner, K., Sommer, U., 2009. Global warming benefits the small in aquatic ecosystems. Proc. Natl. Acad. Sci. USA 106, 12788–12793.

Dawes, C.J., 1998. Marine Botany. John Wiley and Sons Inc., New York.

Dortch, Q., Clayton, J.R., Thoresen, S.S., Ahmed, S.L., 1984. Species differences in accumulation of nitrogen pools in phytoplankton. Mar. Biol. 81, 237–250.

DeMelo, R., France, R., McQueen, D.J., 1992. Biomanipulation: hit or myth? Limnol. Oceanogr. 37, 192–207.

Ducklow, H.W., Steinberg, D.K., Buesseler, K.O., 2001. Upper ocean carbon export and the biological pump. Oceanography. 14, 50–58.

Eby, L., Roach, W., Crowder, L., Stanford, J., 2006. Effects of stocking-up freshwater food webs. Trends Ecol. Evol. 21, 576–584.

Elser, J.J., Chrzanowski, T.H., Sterner, R.W., Mills, K.H., 1998. Stoichiometric constrains on food-web dynamics: a whole lake experiment on the Canadian Shield. Ecosystems 1, 120–136.

Eppley, R.W., 1972. Temperature and phytoplankton growth in the sea. Fish. Bull. 70, 1063–1085.

Fransz, H.G., Colebrook, J.M., Gamble, J.C., Krause, M., 1991. The zooplankton of the North Sea. Neth. J. Sea Res. 28, 1–52.

Frigstad, H., Andersen, T., Hessen, D.O., Jeansson, E., Skogen, M., Naustvoll, L.-J., et al., 2013. Long-term trends in carbon, nutrients and stoichiometry in Norwegian coastal waters: evidence of a regime shift. Prog. Oceanogr. 111, 113–124.

Gilpin, L.C., Davidson, K., Roberts, E., 2004. The influence of changes in nitrogen: silicon ratios on diatom growth dynamics. J. Sea Res. 51, 21–35.

Greenstreet, S.P., Bryant, A.D., Broekhuizen, N., Hall, S.J., Heath, M.R., 1997. Seasonal variation in the consumption of food by fish in the North Sea and implications for food web dynamics. ICES J. Mar. Sci. 54, 243–266.

Hairston, N.G., Smith, F.E., Slobodkin, L.B., 1960. Community structure, population control, and competition. Am. Nat. 94, 421–425.

Hallegraff, G.M., 1993. A review of harmful algal blooms and their apparent global increase. Phycologia 32, 79–99.

Hamre, J., 1980. Biology, exploitation, and management of the northeast Atlantic mackerel. Rapp. P.-v. Réun. Cons. int. Explor. Mer 177, 212–242.

Hansen, B., Bjørnsen, P.K., Hansen, P.J., 1994. The size ratio between planktonic predators and their prey. Limnol. Oceanogr. 39, 395–403.

Heath, M.R., 2007. The consumption of zooplankton by early life stages of fish in the North Sea. ICES J. Mar. Sci. 64, 1650–1663.

Holling, C.S., 1973. Resilience and stability of ecological systems. Annu. Rev. Ecol. Syst. 4, 385–398.

Hudson, J.J., Taylor, W.D., Schindler, D.W., 1999. Planktonic nutrient regeneration and cycling efficiency in temperate lakes. Nature 400, 659–661.

ICES, 2012a. Report of the ICES Advisory Committee 2012. ICES Advice, 2012. Book 6.

ICES, 2012b. Report of the ICES Advisory Committee 2012. ICES Advice, 2012. Book 9.

Jickells, T., 1998. Nutrient biogeochemistry of the coastal zone. Science 281, 217–221.

Johannessen, T., Dahl, E., Falkenhaug, T., Naustvoll, L.J., 2012. Concurrent recruitment failure in gadoids and changes in the plankton community along the Norwegian Skagerrak coast after 2002. ICES J. Mar. Sci. 69, 795–801.

Kagami, M., Urabe, J., 2001. Phytoplankton growth rate as a function of cell size: an experimental test in Lake Biwa. Limnology. 2, 111–117.

Kattner, G., Albers, C., Graeve, M., Schnack-Schiel, S.B., 2003. Fatty acid and alcohol composition of the small polar copepods, *Oithona* and *Oncaea*: indication on feeding modes. Polar Biol. 26, 666–671.

Lee, R.F., Hagen, W., Kattner, G., 2006. Lipid storage in marine zooplankton. Mar. Ecol. Prog. Ser. 307, 273–306.

Lehman, J.T., Scavia, D., 1982. Microscale patchiness of nutrients in plankton communities. Science 216, 729–730.

Leis, J.M., 2007. Behaviour as input for modelling dispersal of fish larvae: behaviour, biogeography, hydrodynamics, ontogeny, physiology and phylogeny meet hydrography. Mar. Ecol. Prog. Ser. 347, 185–193.

Lindahl, O., Hernroth, L., 1983. Phyto-zooplankton community in coastal waters of western Sweden—an ecosystem off balance? Mar. Ecol. Prog. Ser. 10, 119–126.

Lindahl, O., Anderson, L., Belgrano, A., 2009. Primary phytoplankton productivity in the Gullmar Fjord, Sweden. An evaluation of the 1985–2008 time series. Naturvårdverket report 6306, 1–38.

Lynch, M., Sharpiro, J., 1981. Predation, competition, and phytoplankton community structure. Limnol. Oceanogr. 26, 86–102.

Malone, T.C., 1980. Algal size. In: Morris, I. (Ed.), Studies in Ecology. The Physiological Ecology of Phytoplankton. Blackwell, Oxford, pp. 433–463.

Marchall, S.M., Orr, A.P., 1952. On the biology of *Calanus finmarchicus*. Part VII. Factors affecting egg production. J. Mar. Biol. Ass. UK 30, 527–548.

McCarthy, J.J., Goldman, J.C., 1979. Nitrogenous nutrition of marine phytoplankton in nutrient-depleted waters. Science 203, 670–672.

Nielsen, T.G., Sabatini, M., 1996. Role of cyclopoid copepods *Oithona* spp. in North Sea plankton communities. Mar. Ecol. Prog. Ser. 139, 79–93.

North Sea Task Force, 1993. North Sea Subregion 8. Assessment Rep. Norw. State Pollut. Control Authority. 79 p.

Pauly, D., Christensen, V., 1995. Primary production required to sustain global fisheries. Nature 374, 255–257.

Pechenik, J.A., 1999. On the advantages and disadvantages of larval stages in benthic marine invertebrate life cycles. Mar. Ecol. Prog. Ser. 177, 269–297.

Pimm, S.L., 1991. The Balance of Nature? University of Chicago Press, London.

Planque, B., Fromentin, J.-M., 1996. *Calanus* and environment in the eastern North Atlantic. 1. Spatial and temporal patterns of *C. finmarchicus* and *C. helgolandicus*. Mar. Ecol. Prog. Ser. 134, 101–109.

Prokopchuk, I., Sentyabov, E., 2006. Diets of herring, mackerel, and blue whiting in the Norwegian Sea in relation to *Calanus finmarchicus* distribution and temperature conditions. ICES J. Mar. Sci. 63, 117–127.

Reay, P.J., 1970. Synopsis of biological data on North Atlantic sandeels of the genus *Ammodytes*. FAO Fish. Synopsis 82.

Reid, P.C., Edwards, M., Beaugrand, G., Skogen, M., Stevens, D., 2003. Periodic changes in the zooplankton of the North Sea during the twentieth century linked to oceanic inflow. Fish. Oceanogr. 12, 260–269.

Reigstad, M., Wassmann, P., Ratkova, T., Arashkevich, E., Pasternak, A., Øygarden, S., 2000. Comparison of the springtime vertical export of biogenic matter in three northern Norwegian fjords. Mar. Ecol. Prog. Ser. 201, 73–89.

Ricker, W.E., 1954. Stock and recruitment. Can. J. Fish. Aquat. Sci. 11, 559–623.

Riegman, R., Kuipers, R.R., Noordeloos, A.A.M., Witte, H.J., 1993. Size-differential control of phytoplankton and the structure of plankton communities. Neth. J. Sea Res. 31, 255–265.

Robison, B.H., Bailey, T.G., 1981. Sinking rates and dissolution of midwater fish fecal matter. Mar. Biol. 65, 135–142.

Saba, G.K., Steinberg, D.K., 2012. Abundance, composition, and sinking rates of fish fecal pellets in the Santa Barbara channel. Sci. Rep. 2 (716), 1–6.

Sargent, J.R., Falk-Petersen, S., 1988. The lipid biochemistry of calanoid copepods. Hydrobiologia 167/168, 101–114.

Seymour, J.R., Marcos, Stocker, R., 2009. Resource patch formation and exploitation throughout the marine microbial food web. Am. Nat. 173, E15–E29.

Smayda, T., 1990. Novel and nuisance phytoplankton blooms in the sea: evidence for a global epidemic. In: Granèli, E., Sundström, B., Edler, L., Anderson, D.M. (Eds.), Toxic Marine Phytoplankton. Elsevier, New York, pp. 29–40.

Smayda, T.J., 1980. Phytoplankton species succession. In: Morris, I. (Ed.), Studies in Ecology. The Physiological Ecology of Phytoplankton. Blackwell, Oxford, pp. 493–570.

Sommer, U., 1989. Toward a Darwinian ecology of plankton. In: Sommer, U. (Ed.), Plankton Ecology: Succession in Plankton Communities. Springer-Verlag, Berlin, pp. 1–8.

Sommer, U., Gliwicz, Z.M., Lampert, W., Duncan, A., 1986. The PEG-model of seasonal succession of planktonic events in fresh waters. Arch. Hydrobiol. 106, 433–471.

Sundby, S., 2000. Recruitment of Atlantic cod stocks in relation to temperature and advectlon of copepod populations. Sarsia 85, 277–298.

Sætersdal, G., 1980. A review of past managment of some pelagic stocks and its effectiveness. Rapp. P.-v. Réun. Cons. int. Explor. Mer 177, 505–512.

Tang, E.P., 1995. The allometry of algal growth rates. J. Plankton Res. 17, 1325–1335.

Tesch, F.-W., 2003. The Eel. Blackwell Science, Oxford.

Thorson, G., 1950. Reproductive and larval ecology of marine bottom invertebrates. Biol. Rev. 25, 1–45.

Turner, J., 2002. Zooplankton fecal pellets, marine snow and sinking phytoplankton blooms. Aquat. Microb. Ecol. 27, 57–102.

Turner, J.T., 2004. The importance of small planktonic copepods and their roles in pelagic marine food webs. Zool. Stud. 43, 255–266.

Vanni, M.J., Temte, J., 1990. Seasonal patterns of grazing and nutrient limitation of phytoplankton in a eutrophic lake. Limnol. Oceanogr. 35, 697–709.

Vanni, M., Luecke, C., Kitchell, J., Allen, Y., Temte, J., Magnuson, J., 1990. Effects on lower trophic levels of massive fish mortality. Nature 344, 333–335.

Warren, G.J., Evans, M.S., Jude, D.J., Ayers, J.C., 1986. Seasonal variations in copepod size: effects of temperature, food abundance, and vertebrate predation. J. Plankton Res. 8, 841–853.

Weiler, C.S., Eppley, R.W., 1979. Temporal pattern of division in the dinoflagellate genus Ceratium and its application to the determination of growth rate. J. Exp. Mar. Biol. Ecol. 39, 1–24.

Wiborg, K.F., 1940. The production of zooplankton in the Oslofjord 1933–1934. Hvalråd. Skr. Sci. Results Mar. Biol. Res. 21, 1–87.

Wiltshire, K., Malzahn, A., Wirtz, K., Greve, W., Janisch, S., Mangelsdorf, P., et al., 2008. Resilience of North Sea phytoplankton spring bloom dynamics: an analysis of long-term data at Helgoland Roads. Limnol. Oceanogr. 53, 1294–1302.

CHAPTER

8

Variability Enhancing and Variability Dampening Mechanisms in Marine Ecosystems

8.1 INTRODUCTION

8.1.1 Top-Down versus Bottom-up Control in Ecosystems

In 1960, Hairston et al (1960) proposed their famous hypothesis on interactions between trophic levels in terrestrial ecosystems. The essence of their hypothesis is that in a three-level trophic structure the importance of predation versus competition alternates: in lack of predators carnivore populations would become dense and compete for food (herbivores), herbivores would be held below the carrying capacity by their predators (carnivores), which would cause primary producers to become dense and compete for resources ("the world is green"). The paper by Hairston et al. provoked heated debate, and modifications as well as alternative hypotheses were proposed (see Schoener, 1989). The hypothesis by Hairston et al. has gained popularity in limnic biology as the so-called cascade hypothesis, which explains the structure of the pelagic fresh water communities by top-down control (i.e., predation; Carpenter et al., 1985), as opposed to bottom-up control (i.e., resource availability; McQueen et al., 1989).

Negative predator impact on the abundance of prey populations is obviously a basic assumption for "the world is green" and the cascade hypothesis. However, in Chapter 6 I questioned the general validity of this perception. Rather than there being a negative impact of predators on prey populations, I proposed a mechanism for the positive coexistence of

From an Antagonistic to a Synergistic Predator Prey Perspective.
DOI: http://dx.doi.org/10.1016/B978-0-12-417016-2.00008-7

herbivorous zooplankton and their algal prey and other planktonic organisms taking part in the cycling of nutrients, such as bacteria, heterotrophic nanoflagellates (HNF), and viruses. The mechanism implies that phytoplankton, bacteria, and HNF gain competitive advantages by sacrificing part of their clonal populations in order to obtain resources for continuous growth. I call such positive relationship between trophic levels predator-prey synergism (hereafter synergism)—both predator and prey enhance their abundance by coexistence. According to this mechanism the positive coexistence of herbivores and their algal prey depends on growth rates in phytoplankton balancing grazing rates. Ecological experiments commonly using strong doses may therefore be highly deceptive, which may explain why many limnic studies have resulted in apparently negative impacts between phytoplankton and zooplankton (Chapter 7). In Chapter 7, I argued that planktivores too are part of the synergistic interactions in marine ecosystems, whereas their predators (mainly piscivores) will have a negative impact on the abundance of their prey.

Synergism is a top-down force, but in contrast with the classical top-down perspective the relationship between trophic levels is positive. Implications of synergism for the functioning of marine ecology were investigated in Chapter 7. One important impact of synergism is that ecosystems may not respond in a gradual dose-response manner but shift abruptly between alternative stable states. Accordingly, repeated incidents of abrupt and persistent ecosystem shifts (bifurcations) were observed along the Norwegian Skagerrak coast in relation to increasing eutrophication and increasing temperature (Chapters 2, 4, and 5). Synergism in plankton may provide a mechanistic explanation for resilience (sensu Holling, 1973) in marine ecosystems by herbivores stimulating growth of their algal prey and thus dampening the impact of suboptimal environmental conditions (Chapter 6).

8.1.2 Variability in Marine Ecosystems

In spite of the dampening impact of synergism, plankton communities are highly variable on short (e.g., Lindahl and Perissinotto, 1987; Zagami et al., 1996) as well as long (e.g., Lindahl and Hernroth, 1988; Mackas et al., 2001) temporal scales. Interestingly then, high community variability and high resilience are not incompatible phenomena. On the other hand, the dampening impact of synergism will probably induce substantial temporal dependency (autocorrelation) in the plankton community (Chapter 6). This combined with the variability enhancing processes is likely to induce highly complex dynamic patterns in plankton which then will be the basis for energy flow to high trophic levels and affect, among others, recruitment of fish and invertebrates (Chapter 4). Gaining insight into the processes that enhance and those that may counteract variability may thus be important to understand the functioning of marine ecosystems, including how exploitation of marine resources may affect variability and thus the energy flow pattern.

An important motivation for analyzing marine ecosystems from this perspective is related to the high variability observed in marine ecosystems, which, among others, is reflected in inter-annual fluctuations in year-class strength in most fishes (Chapters 2–4; Hjort, 1914;). The question of what causes recruitment variability in marine organisms has been one of the most important but unresolved questions for management of marine resources ever since Hjort (1914) proposed the first recruitment hypothesis 100 years ago. In Chapter 3, I proposed a recruitment hypothesis for Atlantic cod (*Gadus morhua*), which was supported by comprehensive testing in the field (Chapters 3 and 4). The essence of this hypothesis is that the survival of young-of-the-year cod on the south coast of Norway is limited by food availability after settlement (3–5 months old). Good condition and high survival rates in cod were observed in relation to a high proportion of medium to large copepods in the diet, whereas small copepods and hyperbenthic prey such as fish and prawns were associated with poor condition and low survival rates. Since recruitment in cod is highly variable inter-annually, so must the availability of energy-rich planktonic prey. The suggested mechanism underlying recruitment variability was inter-annual variability in the energy flow pattern from lower to higher trophic levels, which in turn is linked to variability in the plankton community structure (Chapter 4). One important question remains: What generates this variability and what could act to dampen variability in marine ecosystems?

This question is investigated under the presumption that synergism is an important structuring mechanism in marine ecosystems, and that variability occurs within a specific stable state. Physical and chemical variability will obviously have a strong impact on the plankton community. Therefore, as a first step in this investigation variability in the phytoplankton community is described on the basis of tri-weekly measurements of algal biomass in terms of Chl *a* and counts of blooming algal species in Flødevigen Bay on the south coast of Norway (Fig. 6.1).

At the end of this chapter, potential evolutionary implications of the ideas presented in this book are briefly reviewed.

8.2 METHODS

Intra- and inter-annual variability in the phytoplankton community was studied on the basis of Chl *a* concentrations at 0–3 m depth measured three times a week in Flødevigen Bay on the south coast of Norway (Fig. 6.1) during the period 1989–2001, a period preceding a likely bifurcation event in these waters around 2002 that had a substantial influence on the algal community (Chapter 5; Frigstad et al., 2013; Johannessen et al., 2012).

To study the inter-annual variability in commencement, duration, and magnitude of algal blooms, a quantitative definition of blooms was developed according to four criteria: (1) Average Chl *a* concentration of five consecutive measurements should peak above 2.5 $\mu g\,l^{-1}$. This limit, which corresponds to the 75th percentile of all measurements between 1989 and 2001, was found to give reasonably good depictions of algal blooms (Fig. 8.1). (2) At least two consecutive measurements during the bloom should be higher than 2.5 $\mu g\,l^{-1}$; hence, high single measurements are not considered as blooms. (3) A bloom is interrupted if Chl *a* concentrations remain below 2.5 $\mu g\,l^{-1}$ for more than a week (three consecutive measurements). (4) The beginning and end of blooms are defined as the dates when the Chl *a* increases above or drops below 2.5 $\mu g\,l^{-1}$.

Blooms were categorized as spring (≤30 April), summer (1 May–14 August) and autumn (≥ 15 August) blooms, in accordance with the dates used in Chapter 5. The dominating algal species during blooms are indicated based on algal counts (methods are described by Dahl and Johannessen, 1999).

Nutrients were measured biweekly at a hydrographical station situated 1 nautical mile offshore, near Flødevigen Bay (Station 201, Fig. 6.1). As an indication of the availability of new (allochtonous) nutrients in the euphotic zone, nitrate concentrations in the upper 10 m of the water column are presented based on measurements at 0, 5, and 10 m depth. Nitrogen is generally the limiting nutrient in these waters, but the spring bloom of diatoms is generally terminated due to the depletion of silicate (Chapter 6). For more details about methodology and position of sampling stations, see Chapter 6.

8.3 VARIABILITY IN PHYTOPLANKTON—RESULTS AND DISCUSSION

The tri-weekly Chl *a* measurements appear as a series of peaks and troughs (Fig. 8.1), both during algal blooms (green areas) and during periods with lower Chl *a*. The spring increase in Chl *a* coincided with the beginning of the spring bloom and the decline in surface water nitrate. The mean start date of the spring bloom and thereby the productive season was 5 March (Table 8.1). However, the onset of the spring outburst varied by almost three months, from 22 January in 1997 to 9 April in 1990. Duration (Table 8.1) and magnitude (Fig. 8.2) of the spring

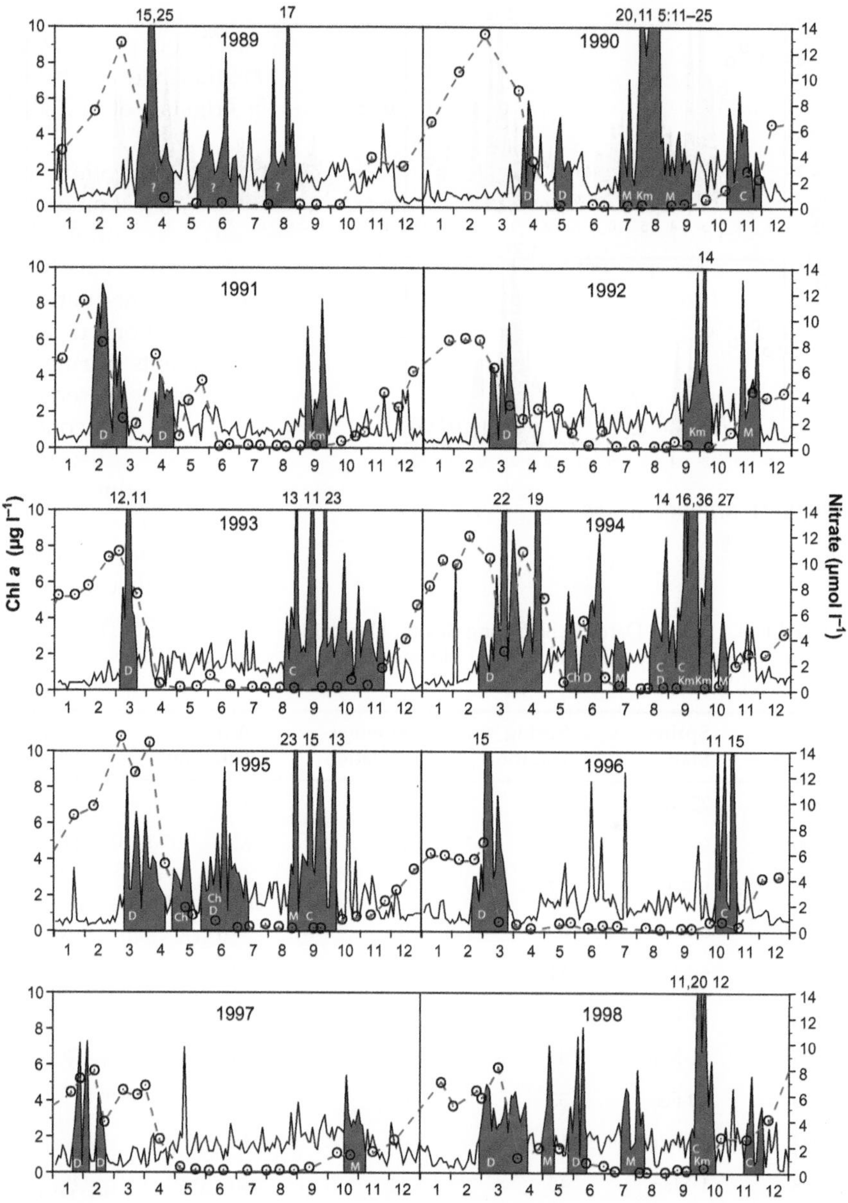

FIGURE 8.1 Measurements of Chl *a* in the upper 3 m in Flødevigen Bay obtained three times a week between 1989 and 2001. Green areas indicate algal blooms. Dominating algae during blooms: C—*Ceratium* spp., Ch—*Chrysochromulina* spp., D—diatoms, Km—*Karenia mikimotoi*, M—mixture of species, ?—not quantified. Broken line indicates nitrate concentrations at 0–10 m depth (integrated from measurements at 0, 5, and 10 m) measured approximately fortnightly.

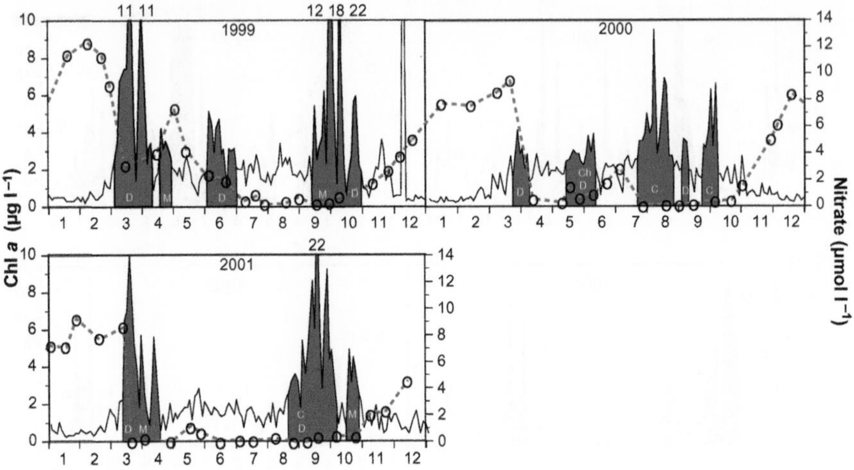

FIGURE 8.1 (*Continued*)

TABLE 8.1 Start Date of the Spring Bloom, Duration in Terms of No. of Days of Spring, Summer, Autumn, and Annual Blooms, and Mean Values of all Years 1989–2001

	Spring Start	Spring Duration	Summer Duration	Autumn Duration	Annual Duration
1989	22-Mar	29	55	7	91
1990	9-Apr	9	41	61	111
1991	5-Feb	52	0	19	71
1992	5-Mar	23	0	44	67
1993	6-Mar	14	0	93	107
1994	27-Feb	59	48	61	168
1995	11-Mar	33	61	43	137
1996	20-Feb	33	0	18	51
1997	22-Jan	18	0	18	36
1998	1-Mar	43	44	30	117
1999	5-Mar	47	26	55	128
2000	24-Mar	8	51	28	87
2001	13-Mar	30	0	52	82
Mean	5-Mar	31	25	41	97

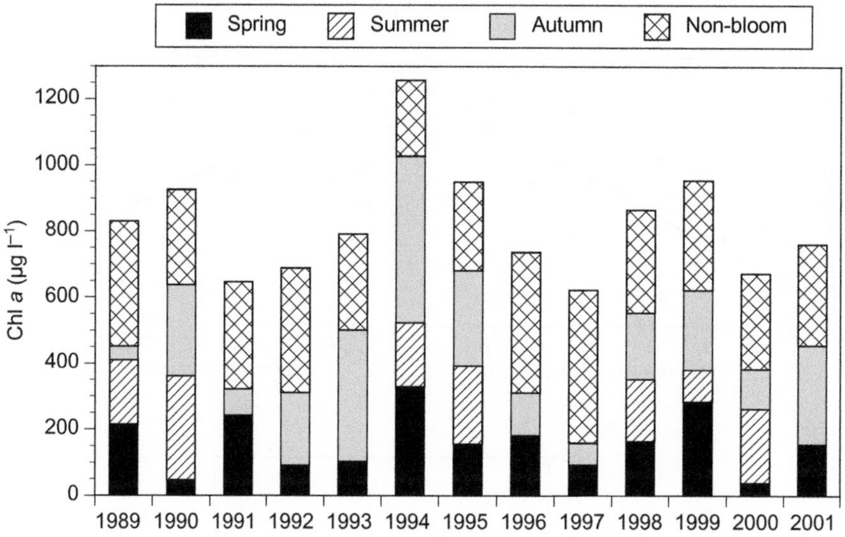

FIGURE 8.2 Integrated Chl *a* during spring, summer, and autumn blooms, and integrated annual Chl *a* during non-blooming periods.

blooms varied substantially too, from 8 days in 2000 to 59 days in 1994. The magnitude of the 1994 bloom was eight times higher than the bloom in 2000. The intensive spring bloom in 1994 was sustained by a pulse of new nutrients in early April, probably from a mixing event that brought nutrients from deeper water into the euphotic layer. In 1991, there was an early spring bloom that was interrupted because of the depletion of silicate (0.46 μmol l^{-1} at 0–10 m; 7 March), and a second bloom occurred in April due to a supply of new nutrients. In general, though, there was one spring bloom that was terminated by the depletion of surface water nutrients (usually silicate; Chapter 6). This also occurred in 1997 ($SiO_2 = 0.78$ μmol l^{-1}; 15 Feb.). However, the early spring bloom that year was unstable, as indicated by a short interruption in early February, and when new nutrients were supplied to the surface water shortly after the bloom, it did not trigger another bloom ($SiO_2 = 3.53$ μmol l^{-1}, 24 Feb.). Nutrients remained high in the surface water until April when the nutrients were depleted without an increase in the algal biomass. Similarly, in 1990 and 2000 when the blooms occurred late in spring, utilization of relatively high concentrations of nutrients only resulted in minimal blooms. This suggests that important productive processes may take place in spring without significantly affecting the phytoplankton biomass. Ratkova et al. (1998) reported similar results from the northern shelf of Norway and suggested that the absence of the spring diatom bloom was due to mesozooplankton grazing.

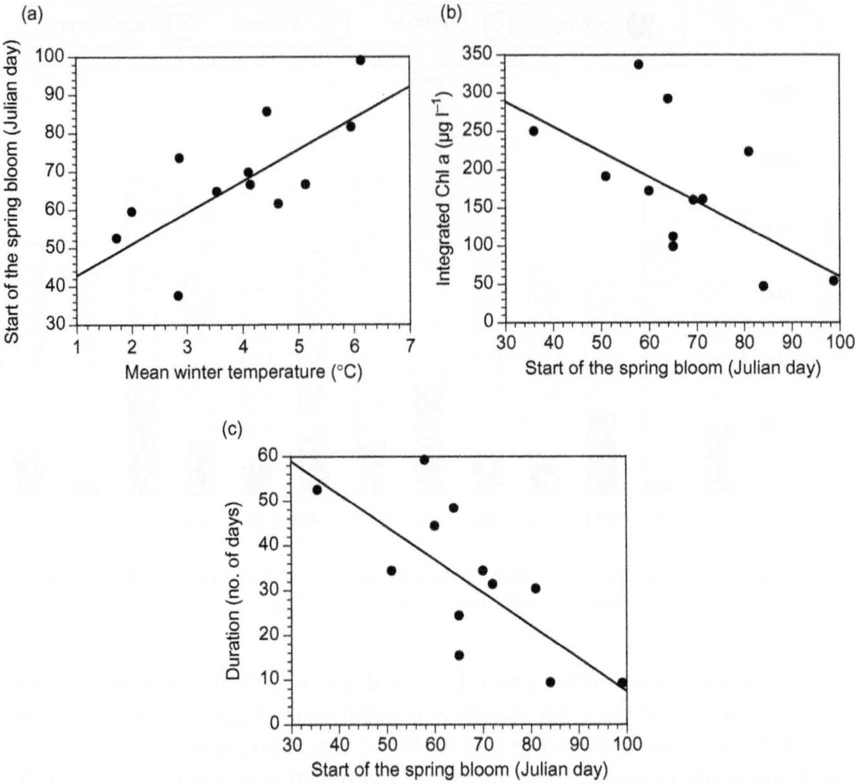

FIGURE 8.3 Spring blooms 1989–2001 (except 1997): (a) start of the spring bloom versus sea surface winter temperature (Feb.–Mar., daily measurements at 1 m depth), (b) magnitude of the spring bloom (integrated Chl *a*) versus start of the spring bloom, (c) duration of the spring bloom (no. of days) versus start of the spring bloom.

Fig. 8.3a shows that there was a positive relationship ($r^2 = 0.53$, $p = 0.008$; linear regression) between the onset of the spring bloom in terms of Julian days and the winter temperature (February and March, from Fig. 5.6b)—that is, in warm years the bloom occurred in late spring (1997 was excluded from all analyses in Fig. 8.3 because it appeared to be an outlier). It is unlikely that there is a direct causal relationship between temperature and the commencement of the spring bloom. On the other hand, higher winter temperatures in these waters are associated with stronger westerly winds (e.g., Greatbatch, 2000), which will generate turbulent mixing. Therefore, the late spring blooms in warm winters support the classical view that the spring bloom is light dependent, which in turn is related to stratification of the water column that prevents phytoplankton from being mixed to deeper and darker water (Sverdrup, 1953).

Early start of the spring outburst was associated with higher magnitude (Fig. 8.3b; $r^2 = 0.35$, $p = 0.043$) and longer duration (Fig. 8.3c; $r^2 = 0.36$, $p = 0.037$) of the spring bloom. Partial correlation analysis between winter temperature, onset of the spring bloom, and duration of the spring bloom gave $r = -0.18$ between duration and temperature, and $r = -0.54$ between the onset and duration, which suggests that an early start of the spring bloom was a more important factor for the duration than low winter temperature. Except for the early bloom in January–February 1997, this year too fits the pattern of low magnitude of short duration of blooms occurring in spring, as there was no marked increase in Chl *a* when nutrients were utilized April that year (1997 was quite warm).

Chl *a* also varied substantially in summer (Fig. 8.2). During the 13-year period, 6 years without summer blooms occurred. When summer blooms did occur, they tended to be quite long-lasting (mean 46.6 days of the 107-day summer period), and in some years the blooms were quite intense (e.g., 1990 and 2000). The blooms in May–June were generally dominated by diatoms, but in some years there were relatively high concentrations of *Chrysochromulina* spp. (Fig. 8.1; Dahl and Johannessen, 1998). *Chrysochromulina polylepis* caused mass mortality in fish and invertebrates in Skagerrak in 1988 (Underdahl et al., 1989), but no mortality was observed during relatively pronounced blooms in 1994 and 1995 (up to 5 million cells l^{-1}). Blooms in late summer often consisted of a mixture of species. However, some of the more intense blooms were dominated by *Karenia mikimotoi* (1990) or *Ceratium* spp. (1995 and 2000). Both *K. mikimotoi*, which is toxic (Yasumoto et al., 1990) and *Ceratium* spp., which are large and therefore not efficiently grazed (Granéli et al., 1989), are well-known red tide species. As in spring, summer pulses of new nutrients were depleted without an increase in the algal biomass (e.g., in 1991 and 1992).

Autumn was generally the most intensive blooming period (Figs. 8.1 and 8.2). However, the blooming varied highly inter-annually from 7 days in 1989 to 93 days in 1993 (Table 8.1), and the magnitude of the 1994 blooms were 12 times higher than in 1989. At the onset of the autumn blooms no new nutrients were available. Rather, nutrients were lost from the euphotic layer at the commencement of the autumn bloom and continued to be lost during the blooms (Figs. 6.2 and 6.3). Hence, the autumn blooms were mainly based on recycled nutrients (Chapter 6). The increase in nitrate in surface water in late autumn was probably linked to remineralization of organic nitrogen compounds.

The autumn blooms were dominated by *K. mikimotoi* and *Ceratium* spp. and occasionally by diatoms. It should be noted that during blooms there were normally a mixture of species so that characterization of the blooms indicates which species or genus dominated. Nevertheless, these characterizations show high inter-annual variability both in terms of which species/genus dominated, the abundances, and when they occurred. For example, the commencement of *K. mikimotoi* blooms varied from late July in 1990 to late September in 1998, and *Ceratium* from mid-July in 2000 to mid-October in 1996. In 1993 the *Ceratium* bloom lasted from mid-August to mid-November (93 days), whereas neither *Ceratium* nor *K. mikimotoi* bloomed in 1997.

Annual Chl *a* varied by a factor of two and displayed the lowest levels in 1991 and 1997 and the highest in 1994 and 1995 (Fig. 8.2). The annual duration of blooms varied from 36 days in 1997 to 168 days in 1994. Accordingly, blooms contributed marginally to annual Chl *a* in 1997 but dominated in 1994.

The results show that irregularity is a predominant feature of the phytoplankton community, with highly variable patterns between years. During the productive season (start of spring bloom to the end of the autumn bloom each year) the correlation between consecutive Chl *a* measurements was relatively low ($r^2 = 0.30$; first-order autocorrelation on log-transformed data), which implies high short-term variability as well. Hydrographically, Skagerrak is highly dynamic because it is a transit area for water masses flowing from the Baltic, Kattegat, and the North Sea to waters farther north (Fig. 6.1). Therefore, one explanation for the high short-term variability in Chl *a* could have been advection of water masses with variable algal concentration. However, Dahl and Johannessen (1998) found high correlations between Chl *a* in Flødevigen Bay and at stations on a transect across Skagerrak to Hirtshals, Denmark (Fig. 6.1), when measurements were obtained the very same day. This indicates relatively high spatial homogeneity. Dahl and Johannessen concluded that advection of water masses with variable algal concentrations could only partly account for high short-term variability in Chl *a* in Flødevigen Bay (i.e., spatial variability), and that a substantial part was temporal variability, that is, algal growth and loss processes (grazing, sedimentation). As short-term variability occurred on relatively large spatial scales, Dahl and Johannessen suggested that meteorological conditions were the main driving force.

Winter concentrations of nitrate in the surface water varied substantially between years, with the highest levels in 1995, when the concentrations were almost three times higher than in 1996 (Fig 8.1). Winter precipitation over northern Europe was high in 1995 and low in 1996 (Aure et al., 1998). Hence, the 1996 concentrations may have been closer to pristine conditions, whereas much of the nitrate in 1995 probably had anthropogenic origin. From the onset of the spring bloom and until July, nitrate concentrations of the surface water varied substantially inter-annually, from being close to zero from March–April onwards in 1993 and 1996 to peaking several times in 1991 and 1994 (Fig. 8.1). This suggests that physical environmental processes that supply nutrients from land and from deeper water (wind, precipitation, winter temperature related to freezing and melting) vary substantially inter-annually and therefore act as variability enhancing mechanisms of the phytoplankton community. Furthermore, physical environmental conditions cause the onset of spring outburst to vary highly inter-annually.

High variability in the phytoplankton community appears to be commonplace in coastal and estuarine habitats (Smayda, 1998), although there can be different patterns between locations mainly due to human impact on coastal ecosystems (Cloern and Jassby, 2009). In summary, bottom-up processes in the form of physical and chemical variability generate high variability in the phytoplankton community inter-annually, seasonally, and within days.

8.4 IMPACT OF HERBIVOROUS ZOOPLANKTON ON PHYTOPLANKTON VARIABILITY

The mechanism behind synergism in plankton is based on the principle that herbivorous zooplankton stimulate the production of their preferred algal prey and thus have a strong impact on the phytoplankton community. An important consequence of this is that zooplankton will dampen the impacts of suboptimal environmental conditions for their algal prey. Hence, physical and chemical variability generates variability in the plankton community, whereas synergistic herbivores dampen the impact of these bottom-up processes. A likely consequence of synergism in plankton is that there will be no simple dose-response relationship between environmental and biological variables. The complexity of disentangling the impacts of environmental perturbations or gradual environmental changes in systems with synergistic plankton can be illustrated by the seasonal succession in the plankton community.

In a synergistic system, the level of primary productivity is an essential structuring mechanism for the plankton community (see Chapter 7 for detailed description). In the coastal waters of Skagerrak, the annual pattern in primary productivity is dome-shaped with minimum production in winter and maximum in summer (Fig. 7.1). Relatively low primary productivity in spring and autumn favor large synergistic plankton, and high primary productivity in summer favors small synergists. The explanation for the increasing primary productivity from spring to summer is a "natural eutrophication process" caused by nutrients from deeper water being mixed gradually into the euphotic layer, were the nutrients are captured and recycled in the plankton community. The over-wintering stocks of herbivores are the memories of the system (Chapter 6), and the most abundant species will initially have the highest impact on the seasonal succession by modifying the algal community in favor of themselves. To become abundant, the herbivores depend on successful recruitment. The most abundant adult stocks will have the highest number of offspring, which in turn may modify their microhabitat in favor of themselves and their algal prey. However, there are two factors that may counteract this predictable impact of synergism, namely the onset of the productive season and the duration and magnitude of the spring bloom of diatoms.

The spring bloom of diatoms has been considered particularly important for the great fisheries (Cushing, 1989). However, as a basis for energy flow into the pelagic food web there is growing evidence to suggest that the spring bloom has been substantially overrated. For example, the spring period (February–April) contributed merely 20% to the annual primary production in Skagerrak (Fig. 6.2). Sedimentation is high during the spring bloom (Smetacek, 1985; Turner, 2002; Wassmann, 1991) and therefore probably important for energy supply to the benthic food web (e.g., Townsend and Cammen, 1988), and some diatoms may have a negative impact of the reproductive success of copepods (Ianora et al., 2003; Pierson et al., 2005; Wichard et al., 2005). Diatoms are non-motile in general and therefore not able to exploit microscale nutrient patches from zooplankton exudation and excretion efficiently. Motility and the ability to utilize microscale resources in phytoplankton, bacteria, and HNF are important factors for synergism in plankton (Chapter 6). Diatoms are thus non-synergistic plankton that will be negatively affected by grazing, which may explain why some diatoms have developed grazing deterrents (Pondaven et al., 2007).

Despite the likely overrating of the significance of the spring bloom for the pelagic community, there is substantial evidence to suggest that copepods graze diatoms (e.g., Irigoien et al., 2002; Koski, 2007). In agreement with this, grazing was probably the reason for the short and low-magnitude blooms when the productive season started late in the spring, or the depletion of nutrients without increase in the algal biomass (Figs. 8.1 and 8.2). The onset, duration, and magnitude of the spring bloom varied highly inter-annually. This will obviously have consequences for the seasonal succession, for example, by inducing match or mismatch between herbivores and the spring bloom. Temporal mismatches may result in substantial sedimentation of diatoms and thereby loss of adequate food for their main grazer, relatively large herbivores (Reigstad et al., 2000). The duration and magnitude of the spring bloom are affected by the onset of the productive season (Fig. 8.2), by mixing and upwelling events bringing new nutrients into the euphotic layer (e.g., 1991, 1994, and 1999; Fig. 8.1) and by nutrients being flushed into the sea in relation to variable winter temperatures and precipitation (rain or snow). All these factors that affect spring bloom dynamics may be likened to climatic variability and physical forcing. However, the outcome of a series of physical and chemical events during spring depends on the composition and overall abundance of overwintering herbivores (e.g., Sommer et al., 2012), which affect both the capacity to graze the blooming diatoms and how the herbivores will shape the synergistic part of the phytoplankton community. Synergistic phytoplankton will not bloom but are likely to be present also during blooms of non-synergistic phytoplankton. After the spring bloom, the algal community becomes controlled by synergistic herbivores that are able to dampen negative impacts for their algal prey. The seasonal succession toward smaller synergists will be driven by the rate of increase in primary productivity (Chapter 7), which in turn is linked to physical processes (wind mixing, upwelling, precipitation, snow melting) that contribute to the euphotic layer's nutrient supply. A series of early wind-mixing or upwelling events will thus result in a rapid increase in primary productivity, whereas calm conditions will cause primary productivity to increase more slowly and thereby creates a longer temporal window for larger herbivorous zooplankton and their algal prey. The rate of change in primary productivity will also be affected by the abundance of planktivores feeding on the herbivores. This is related to the planktivores incorporating nutrients into somatic growth and by causing increased export of nutrients out of the euphotic layer (Chapter 7). Large stocks of planktivorous fishes will thus contribute to lower primary productivity and a longer temporal window for larger synergistic plankton.

In summary, the plankton community structure during the seasonal succession will be a result of physical and chemical variability, abundance of overwintering stocks of herbivores, and the damping impact of synergism in plankton. Synergism will thus result in substantial temporal dependency (autocorrelation). To predict the impact of physical perturbations on the plankton community requires detailed knowledge of the plankton community structure just prior to the perturbation. Without such knowledge, studies of the relationship between physical variables and responses in the plankton community are probably in most cases doomed to fail.

8.5 RECRUITMENT OF FISH AND INVERTEBRATES

8.5.1 Variable Energy Flow Patterns

Ever since Hjort (1914) proposed the first recruitment hypothesis for fish 100 years ago, one of the main candidates for the underlying mechanism has been physical variability. This has sparked numerous correlation studies that over the years have become increasingly complex in terms of statistical methods. Many significant correlations have been published (e.g., Shepherd et al., 1984), but as most of us tend to publish mainly positive findings, we can only guess how many unpublished correlations there were for each that was published. Without taking this number into account, the danger of committing a type I statistical error increases substantially with the number of correlations (type I error is the incorrect rejection of a true null hypothesis). Hence, there are probably numerous spurious correlations that have been published. Indeed, if there had been simple dose-response relationships in marine ecosystems, and the physical environmental conditions are important for recruitment variability, the recruitment puzzle would have been solved a long time ago.

Most "significant" correlations have not stood the test of time (Myers, 1998), and more-modern and sophisticated statistical methods which can handle numerous variables that are inter-correlated (e.g., Johannessen and Tveite, 1989) have not improved this dismal situation. The only correlations that seem to be fairly robust are, not surprisingly, for stocks that are at the limit of their geographical range (Myers, 1998).

On the other hand, there are obvious reasons for searching for physical-biological coupling as driving forces behind recruitment variability. First, a specific spawning stock biomass (or biomass of spawning products) with a specific age distribution (which relates to maternal effects, e.g., Green, 2008) may give rise to very weak and very strong year-classes (e.g., Rothschild, 1986; Shepherd and Cushing, 1980). Second, there is general consensus that most of the recruitment variability occurs during early life stages in most fishes (Chapter 3). Third, the variability in the plankton community on short (e.g., Lindahl and Perissinotto, 1987; Zagami et al., 1996) as well as long (e.g., Lindahl and Hernroth, 1988; Mackas et al., 2001) temporal scales appears to be sufficiently high to match that of recruitment. Also, as the primary producers are microorganisms that can respond extremely rapidly to favorable conditions, physical and chemical variability stand out as the main driving forces behind variability in the plankton community.

The critical period for recruitment in cod along the Norwegian Skagerrak coast is in summer, after cod have settled from the pelagic to the benthic habitat (Chapters 3 and 4). Despite that the cod were relatively large during the critical period (3–9 cm), survival was related to the proportion of energy-rich planktonic prey in their diet, with relatively large herbivorous copepods being particular favorable. Physical and chemical variability generates variability in the plankton community, whereas synergistic interactions in plankton act to dampen variability. The process leading to variable abundance of adequate planktonic prey for 0-group cod during the critical period in summer will thus be a complex mixture of bottom-up and top-down processes, starting with the composition of overwintering herbivores, the concentrations of nutrients in winter, onset and duration of the spring bloom, and the pattern of random events that mixes nutrients into the euphotic layer.

In addition, variability in the advection of water masses with different physical and biological characteristic adds to this complexity. Therefore, in spite of the bottom-up processes probably being the main driving forces behind the recruitment variability, the chance of finding correlations between recruitment in cod and physical and chemical variables are meager. An exception might be the decrease in the size of herbivores with increasing primary productivity. Low mixing of nutrients into the euphotic layer in spring to early summer will open for a longer temporal window for larger herbivores, which will be beneficial for the survival of cod during the critical period. Interestingly, Johannessen and Tveite (1989) found that stable temperature in subsurface water (19 m depth) in spring was positively correlated with cod recruitment along the south coast of Norway. As the temperature in turn was related to wind conditions, the authors took this as an indication of low mixing of the water column.

Variable energy flow patterns in the plankton community will probably have a strong impact on the survival of most organisms that depend on planktonic prey during early life stages, which is the vast majority of teleost fishes (Leis, 2007) and a high proportion of benthic invertebrates (Pechenik, 1999; Thorson, 1950). In contrast to recruitment hypotheses that focus on match-mismatch between larval abundance and the spring and autumn outbursts (Cushing, 1975; Hjort, 1914), variable energy flow patterns will affect the survival of young stages of planktivores throughout the productive season. In agreement with this, high variability of the abundance of young-of-the-year spring spawning as well as summer-spawning fishes was evident along the Norwegian Skagerrak coast (Chapters 3 and 4). Furthermore, in his seminal paper, Thorson (1950) concluded that the variability in benthic invertebrates with a long pelagic larval phase was generally much higher than in invertebrates with a short pelagic phase or without pelagic larvae.

Although this energy flow hypothesis was developed and tested for cod along the Norwegian Skagerrak coast, the underlying mechanism of variable energy flow pattern will probably have general validity for marine spring-blooming ecosystems. And, as planktonic organisms may respond rapidly to favorable environmental conditions, variable energy flow patterns may potentially be a common characteristic of aquatic pelagic food webs, with the degree of variability depending on the intensity of physical forcing. The dependency on energy-rich pelagic prey may affect survival in planktivores well beyond the larval stage, probably until the organism has become sufficiently large to withstand longer periods without energy-rich pelagic prey or large enough to shift to and persist on less energy-rich prey. For example, for 0-group cod the critical period for recruitment was over in August in relation to the explosion in the abundance of young-of-the-year fish and invertebrates that spawn in summer (Chapters 3 and 4). The size at which organisms grow beyond the critical stage of food-limited survival will probably vary between species.

8.5.2 Recruitment Hypotheses

Hjort's (1914) recruitment hypothesis postulates that the year-class strength in the Arcto-Norwegian cod is determined at the early larval stage, depending on the degree of match between the abundance of larvae and planktonic prey. The tests of recruitment in cod presented here do not support Hjort's hypothesis. Cushing (1975, 1990) refined Hjort's hypothesis and called it the match-mismatch hypothesis. This hypothesis is inextricably linked to the spring and autumn outbursts, as underlined by Cushing (1990, p. 284): "There are three consequences of the match/mismatch hypothesis. The first is that fish should release their larvae during the spring and autumn peaks in plankton production in temperate waters ....The second consequence of the match/mismatch hypothesis is that correlations (if necessarily low) between larval numbers and recruitment are expected." Neither prediction is supported by the results presented in this book. The number of cod that settled (i.e., after the larval phase and therefore expected to improve the correlation) was poorly correlated with recruitment ($r^2 = 0.16$, abundance in July versus September), whereas the abundance of cod after the critical phase corresponded well with that of recruitment ($r^2 = 0.96$, August versus September, recruitment is mainly determined by September in these waters; Chapter 4). Furthermore, the majority of shallow-water fishes along the Norwegian Skagerrak coast spawn in summer, and they are far more successful than the spring spawners both in terms of number and biomass of young-of-the year fish (Chapters 3 and 4).

The summer spawners are mainly small annual gobies (approximately 4–5 cm as adults). The abundance was drastically reduced during winter (in one instance, the abundance of two-spot goby [*Gobiusculus flavenscens*] was reduced from 42,000 in September to 13 per beach seine haul in June the next year (Chapter 3). In the spring spawning gadoids, there was good correspondence between the abundance at the 0- and I-group stages ($r^2 = 0.72$ and 0.69 for cod and pollack (*Pollachius pollachius*) respectively; Chapter 2), which suggests that survival through the winter was relatively stable. In addition, the mortality during winter was lower in the gadoids than in the annual gobies. The spring spawning gadoids are much larger than the gobies in late autumn. Spawning in spring may thus be a strategy in larger and more long-lived fishes that reduces the mortality during the first winter. As commercial fishes are generally much larger than the annual gobies, the perception that fish tend to spawn in spring may therefore by biased by fisheries biologists focusing on commercial species.

Durant et al. (2007) extended the match-mismatch hypothesis to include the role of climate in affecting the reproductive success of a predator through its effect on the relative timing of food requirement and food availability during early life stages in general. In this general form the match-mismatch hypothesis comprises the timing aspect of variable energy flow pattern; for example, a short temporal window for larger herbivores may result in a mismatch with the critical period for recruitment of cod. However, in contrast to the match-mismatch hypothesis that focuses on temporal aspects, the energy flow hypothesis focuses on variability in the plankton community structure as a mechanism that affects survival, in which the temporal aspect is only a part of the mechanism.

Food-limited survival during early life stages in fish has often been coupled with reduced predation rates in relation to increasing growth rates when food conditions are good (e.g., Shepherd and Cushing, 1980; Ware, 1975), which in turn is related to the perception that "bigger is better" (e.g., Sogard, 1997). However, during the critical period for cod along the Norwegian Skagerrak coast there was no evidence suggesting that larger cod had higher survival rates than smaller cod. In fact, there were incidents when smaller cod had higher survival rates than larger cod. Therefore, it was suggested that there is an optimal size between cod and available prey that maximize survival (Chapter 4). In this perspective, growth in cod can be viewed as an adaptation to the prevailing food conditions in these waters. Interestingly, there is genetic evidence of separate fjord and archipelago populations in cod (Knutsen et al., 2010), which have different growth rates (Dannevig, 1949). It was suggested that adaptation to different size spectra in planktonic prey could be a mechanism for the genetic separation of cod over relatively short distances (Chapter 4). Also, between the gadoid species the growth rates varied substantially during early life stages (Chapter 3). Hence, it appears that the growth rates in fish during early life stages are adaptations to prey conditions that maximize survival rather than rapid growth in order to minimize predation. However, this does not necessarily imply adaptations to "average conditions." For long-lived species in particular, adaptations to more rare events that give rise to extraordinary high survival rates may result in more offspring during the lifetime of an organism than adaptation to average conditions. One species that might have obtained this strategy is haddock (*Melanogrammus aeglefinus*), which occasionally produce extremely strong year-classes (Fogarty, 2001). Species with this strategy are obviously vulnerable to recruitment overfishing. More short-lived species might be more adapted to average conditions. In agreement with this, there is a general increase in recruitment variability with longevity (Longhurst, 2002).

During the last three decades, there has been growing interest in the maternal impact on successful recruitment in marine organisms both at the individual as well as at the population level (Green, 2008). It has been suggested that age truncation commonly induced by fisheries may have severe consequences for long-term sustainability of fish populations (Berkeley et al., 2004; Birkeland and Dayton, 2005). The latter implies that there is a maternal impact on the overall recruitment of fish stocks. Obviously, as the recruitment puzzle has not been resolved, this remains an untested hypothesis. In a comprehensive review of maternal impacts on fish populations, Green (2008, p. 36) states: "The extrapolation of laboratory results from a number of small individuals to entire populations, and from one trait to recruitment are frequent misrepresentations of the expression and importance of maternal effects in fishes." There are a number of potential maternal impacts on offspring, but the well-documented positive relationship between female size and egg size has received most of the attention: Larger eggs result in larger and more viable larvae. For this to have an impact on recruitment, the early larval stage must be critical. Obviously, it is highly unlikely that there is a maternal impact on the recruitment of cod along the Norwegian Skagerrak as the critical period is in summer, well beyond the larval stage (Chapters 3 and 4). Theoretically though, if the spawning stock should become so low that survival during early larval stages limits recruitment there could be a maternal impact. Whether this may occur in unexploited stocks and thus exert a selection pressure on an adequate evolutionary time scale seems doubtful.

Ware (1975) studied fish recruitment from a theoretical perspective and concluded that it would be beneficial for fish to produce progressively smaller eggs from spring to summer and larger eggs from summer to autumn. This is related to incubation time of the eggs. There may also be another advantage of this pattern in egg size, as the size in zooplankton follows the same pattern from relatively large zooplankton in spring to small in summer and larger again in autumn (Beaugrand et al., 2003; Chapter 7). In agreement with this pattern, larger cod spawn earlier in spring than small cod, and the size of the eggs decreases with successive spawning batches in the individual females (Chambers, 1996; Kjesbu, 1989). This pattern is also observed in other fishes (Green, 2008). As predicted, in the autumn this pattern is reversed with progressively larger eggs in fish toward winter (Chambers, 1997). The change in egg size during the seasonal succession may thus be an adaptation to the seasonal change in zooplankton and the benefits of shorter incubation period in summer. The perception that large eggs in general are beneficial is in conflict with the evolutionary process selecting for optimal size in eggs and larvae. Therefore, in an evolutionary perspective it seems more likely that increase in egg size with age and size in females is a life cycle strategy that provides larger individuals with intra-specific competitive benefits, for example, by outcompeting smaller individuals on the spawning grounds in early spring. It is important to keep in mind that fecundity (number of eggs) in marine organisms is generally very high, so that if only two of the millions of eggs a female cod spawn during her lifetime survive to adulthood, it is sufficient to sustain the stock.

8.6 IMPACT OF PREDATION, PARASITISM, AND DISEASES ON VARIABILITY IN MARINE ECOSYSTEMS

8.6.1 Predation

The studies of recruitment in cod along the Skagerrak coast provided substantial evidence that survival and thereby year-class strength was related to a diet of relatively large and energy-rich copepods, whereas there was no relationship between the survival in cod during the critical period and the abundance of potential predators (Chapters 3 and 4). Hence, recruitment in cod is density-dependent, with availability of high-quality food as the limiting factor. After the critical period, there is good correspondence between year-classes (Chapter 2; Tveite, 1971), that is, a strong year-class remains strong throughout most of their lifetime and weak year-classes remain weak (but see conditions after 2002, Chapter 5). This suggests that neither food nor space sets limits for the abundance of cod after the critical recruitment phase. In agreement with this, there is massive evidence that in fishes there is room for very strong year-classes at the adult stage, for example, the legendary 1904 year-classes of Arcto-Norwegian cod and Norwegian spring-spawning herring (*Clupea harengus*; Hjort, 1914). In long-lived fishes extremely strong year-classes can dominate landings over many years, such as the 1950 year-class of the Norwegian spring-spawning herring that contributed substantially to the landings as late as 1965 (Toresen and Østvedt, 2000). Interestingly, the herring spawning stock that gave rise to the extremely strong year-class in 1950 (the highest on record) was estimated to about 13 million tons, which then constituted the largest single fish stock unit in the North Atlantic (Dragesund et al., 1997). Hence, even in a huge fish stock at the peak of its abundance there is apparently no resource limitation (food and space) for extremely strong year-classes once they have passed through the critical recruitment phase.

Consequently, it appears that survival in fish and invertebrates that feed on plankton during early life stages is density-dependent until having become sufficiently large to withstand periods without adequate planktonic prey or large enough to feed on alternative, less energy-rich prey. After the critical recruitment phase, resources no longer limit survival (except for some benthic organisms that may be limited by both food and space). At this stage predation becomes an important regulating factor. Theoretical studies have shown that specialist predators may drive a predator-prey limit cycle (e.g., May, 1981). In contrast, the functional response to generalist predators tends to be stabilizing (e.g., Hanski et al., 1991). To my knowledge there are not many piscivorous specialists, while abundant predators such the gadoids prey on a wide range of fishes and invertebrates (e.g., Jiang and Jørgensen, 1996; Michalsen et al., 2008; Sarno et al., 1994). Also, huge pelagic fish stocks of herring, mackerel (*Scomber scombrus*), and blue whiting (*Micromesistius poutassou*) in the Norwegian Sea feed on variety of planktonic prey, including fish larvae (Prokopchuk and Sentyabov, 2006). Predation as a general phenomenon will thus act to dampen variability. For example, strong year-classes of cod at the 0-group stage will probably attract more opportunistic predators than weak year-classes, thereby reducing the differences between strong and weak year-classes as the cod grow older. In agreement with this, based on a comprehensive study Myers and Cadigan (1993) reported that differences at the 0-group stage of various fishes were significantly reduced at older stages. Consequently, predation will dampen recruitment variability generated by variable energy flow pattern.

8.6.2 Parasitism and Diseases

There is very little knowledge of the impact of parasitism and pathogens on the abundance of marine organisms. However, in general it seems likely that both parasites and pathogens will be more effectively spread at high host abundances than at low. Hence, parasitism and diseases probably contribute to dampen variability in marine ecosystems.

8.7 IMPACT OF FISHING ON FISH RECRUITMENT AND ECOSYSTEM VARIABILITY

The evaluation of the potential impact of fishing on recruitment does not include recruitment overfishing, but how fishing might impact the structure of the plankton community and thereby affect recruitment in fish, including altered predation pressure on recruiting fish. An important effect of human exploitation of marine resources follows from the arguments that predation on post-recruits will generally reduce the difference between strong and weak year-classes. Human exploitation is typically targeting larger fish and sea mammals. Consequently, predator stocks will be reduced, and the general impact of reduced predation will be increased ecosystem variability (Cury et al., 2003). Theoretically then, recruitment will increase due to reduced predation and thus operate as a compensatory mechanism for the population (Rose et al., 2001). This is probably true when human exploitation is targeting piscivores or benthivores (feeding on fish or benthic invertebrates). However, exploitation of planktivorous fishes may result in a more complex scenario due to the impact planktivores may have on the plankton community: indirectly by affecting the structure of the plankton community through changes in primary productivity (Chapter 7), and directly as predators of early life stages of fish.

According to the synergism hypothesis, large stocks of planktivores will have a positive impact on the abundance of their main prey, large herbivores (Chapter 7). As herbivorous zooplankton have higher nutritional value than larval fish and carnivorous planktonic prey (Chapter 4; Lee et al., 2006; Sargent and Falk-Petersen, 1988), it seems reasonable to assume that planktivores will feed preferentially on large herbivores. In the case of overexploitation of planktivorous fishes, the abundance of large herbivores will be reduced and the temporal window for larger herbivore will become shorter both in spring and autumn, thus resulting in an extended period in summer with low abundance of larger herbivores. The outcome of this is difficult to predict. Assuming fish are adapted to a particular energy flow pattern in the plankton community, altered energy flow may have a negative impact on survival through the critical recruitment phase. This was suggested by the repeated incidents of recruitment collapses in gadoids in relation to abrupt shifts in the plankton community (Chapters 2 and 5). Also, for fish and invertebrates whose survival during the critical recruitment phase depends on relative large and energy-rich planktonic prey during summer, there is likely to be negative impact on recruitment, whereas those adapted to smaller plankton may experience increased recruitment.

Reduced stocks of planktivores will result in reduced predation on plankton. However, the outcome of this is difficult to predict. Reduced abundance of large herbivores will probably result in a shift in the diet of the planktivores to a higher proportion of carnivorous prey because the small herbivores may not provide sufficient energy for larger planktivores, like herring and mackerel (Chapter 7). Due to prolonged periods with low abundance of large herbivores, species that have their early life stage during summer may experience increased predation, whereas those that have early life stages in spring and autumn may experience reduced predation and thereby increased recruitment.

In summary, human exploitation of marine resources will result in reduced predation, and the general impact will be increased variability in marine ecosystems. Fishing targeting piscivores and benthivores will probably result in a compensatory increase in the recruitment of fish and invertebrates, whereas reduced stocks of planktivores may have a negative impact on the recruitment of some species but positive impact on other species.

8.8 QUASI-STABLE STATES IN MARINE ECOSYSTEMS

Synergism in plankton provides a mechanistic explanation for the lack of simple dose-response relationships, resilience, and the vulnerability of marine ecosystems to bifurcations (Chapter 6). Synergists are thus able to persist under suboptimal conditions. High variability in the plankton community is apparently not in conflict with high resilience. Hence, resilience is not a result of homeostatic forces that pull the system back into balance after a perturbation. Rather, the combination of high variability and high resilience suggests that within a specific ecosystem state, the diversity in synergistic plankton is sufficiently high to fill niches as they appear. In the case of extraordinary favorable environmental conditions, synergistic herbivores may become highly abundant, and because of their ability to dampen impacts of negative environmental conditions they may dominate the system longer than the environmental conditions should indicate. Hence, within a specific stable state there may be quasi-stable states that may persist temporarily due the dampening impact of synergism (i.e., the same forces that cause ecosystems to persist within stable states).

One example of quasi-stable states could be the "regime shifts" in the North Pacific (e.g., Hare and Mantua, 2000). According to Overland et al. (2008, p. 100), "There is no convincing evidence that these climate shifts in the North Pacific are between multiple stable states. Instead, they appear to be more consistent with a long memory process with considerable autocorrelation at multi-year time scales, which can show persistent major deviations from a single century scale mean." Another example could be one in the Northeast Atlantic, where the huge blue whiting stock (*Micromesistius poutassou*) experienced unprecedented recruitment over a period of ten years (1995–2004; ICES, 2012).

There are various conditions that may cause ecosystems to persist in quasi-stable states. As mentioned, extraordinary favorable environmental conditions may result in the dominance of a group of synergistic herbivores. Another mechanism could be extraordinary good conditions for recruitment for a large planktivorous fish stock, for instance as happened in 1950 in Norwegian spring-spawning herring (Toresen and Østvedt, 2000). If not subjected to intensive fishing (like the herring was), and the fish is relatively long-lived, the biomass of the stock will increase substantially and remain high over many years. According to the synergism hypothesis, this will favor large herbivores due to reduced primary productivity (Chapter 7). The positive impact of planktivores on larger herbivores in combination with the ability of synergistic herbivores to persist under suboptimal environmental conditions may result in a relatively resilient quasi-stable state.

Conversely, overfishing of planktivores may also result in a relatively resilient quasi-stable state, but in a situation beneficial to smaller herbivores. The collapse of the huge stocks of mackerel and herring in the North Sea in the late 1960s from overfishing, which was accompanied by substantial reduction in the abundance of the relatively large copepod, *C. finmarchicus* (see details in Chapter 7), might be an example of a fishery-induced shift between quasi-stable states.

In summary, synergism will cause substantial autocorrelations in marine plankton communities. This, in combination with bottom-up processes that generate high variability in plankton, will probably generate complex temporal patterns in the plankton community. The potential existence of quasi-stable states exacerbates this complexity. The general failure of correlation studies of recruitment variability and environmental variables (Myers, 1998; Shepherd et al., 1984) is strong evidence in support of such complexity.

8.9 IMPACT OF PELAGIC PROCESSES ON BENTHIC COMMUNITIES

The main focus of this book is the pelagic realm because that is where the important processes occur in relation to the results reported from the long time series that forms the basis of the book. However, processes in pelagic food webs have substantial impact on benthic communities. Here, some of these processes are briefly reviewed. It is beyond the scope of this section to analyze processes within the benthic community.

As mentioned, an important impact on the benthic community is variable recruitment in benthos with pelagic larvae, which was suggested to be a result of variable energy flow patterns in the plankton community (Chapter 5). Thorson (1950) found that abundance in benthic invertebrates with long pelagic stages is much more variable than in invertebrates with short larval stages or without pelagic larvae.

Primary production can only occur in the surface layer (generally less than 100 m) of the oceans due to light limitations in deeper water. Consequently, except for in very shallow seas and otherwise in a relatively narrow belt along the shore where marcoalgae, seagrasses, and benthic microalgae build up high biomasses, benthic habitats depend on organic matter produced by planktonic autotrophs in the euphotic surface layer of the oceans. Export from the euphotic layer is tightly coupled with the structure of the pelagic community. For example, sedimentation of organic matter is high in relation to algal blooms and low when the algal community is under grazing control (Turner, 2002; Wassmann, 1991). Bottom-up processes that generate variability in the plankton community, for instance by causing a mismatch between the spring bloom and the occurrence of herbivores, will thus have an impact on the export of organic matter and act as a variability enhancing mechanism in the benthic habitat as well. Physical environmental variability that affects the export of organic matter from the euphotic layer is a typical stochastic process. In contrast, fishing and other human impacts (e.g., eutrophication) on the pelagic food web that may affect export of organic matter to the benthic food web are more predictable. Pelagic fish contribute directly to the export of nutrients by producing large, fast-sinking fecal pellets from which few nutrients are lost during the rapid descent to the bottom (Robison and Bailey, 1981; Saba and Steinberg, 2012). Due to the high abundance of fish in many oceanic environments and the high nutrient content of their fecal pellets, fish fecal matter may be a major source of organic export from the pelagic community to the benthos (Staresinic et al., 1983; Turner, 2002). In addition, planktivorous fishes may have an indirect impact on the export of organic matter by affecting the size of herbivorous zooplankton. In general, sedimentation of fecal pellets from small copepods is negligible (Turner, 2002), whereas fecal pellets from larger zooplankton can be substantial: Wassmann et al. (1998) reported that, on average, 11–37 % of particulate organic vertical carbon flux was comprised of fecal pellets on the shelf bank off northern Norway. Consequently, overfishing of planktivorous fishes may result in a shift to smaller herbivores, which in turn may result in reduced export of organic matter to the benthos.

The oceans are important sinks of anthropogenic CO_2 (Sabine et al., 2004), and the biological pump (Ducklow et al., 2001) is one of the mechanisms that contributes to export CO_2 from the surface to deeper waters. According to the synergism hypothesis, global warming may induce abrupt shifts in the plankton community (Chapter 7). This in turn may cause recruitment failure in fishes, the collapse of fish stocks, and a shift toward smaller plankton. Consequently, both the direct impact of the collapse of planktivorous fishes stocks and the indirect effect resulting in smaller plankton may reduce the efficiency of the biological pump and thus exacerbate the accumulation of greenhouse gases in the atmosphere.

8.10 EVOLUTIONARY IMPLICATIONS OF PREDATOR-PREY SYNERGISM

Mutations and natural selection are cornerstones of evolution (Ridley, 2003). Mutation rates depend on the enzymes that replicate

DNA and correct errors, and, since the enzymes are produced by genes, the mutation rates are under genetic control (Maynard Smith, 1977). Accordingly, there is evidence of adaptive evolution of mutation rates (Metzgar and Wills, 2000). Being under genetic control, it seems likely that there will be some individual differences in mutation rates between genes coding for similar traits and between genes coding for different traits, although these differences are probably very small. It was argued that synergisms between phytoplankton and grazers and between grasses and their grazers are evolutionary stable strategies (Chapter 6), that is, it cannot be invaded by mutations resulting in non-palatability. Hence, there is probably a selection pressure toward reduced mutation rates due to slightly lower reproductive success in individuals with higher mutation rates for non-palatability and other traits that are incompatible with the synergetic coexistence of predator and prey. Interestingly, by using evolutionary game theory, Bergstom and Lachmann (2003) found that contrary to the Red Queen theory, in mutualism evolution the slowly evolving species is likely to gain a disproportionate share of the benefits. Synergism resembles mutualism by being a result of positive coevolution between species.

Consequently, in ecosystems where synergism plays an important role, mutation rates likely decrease over time and eventually become very low. However, higher mutation rates are probably beneficial during evolution of synergism. Hence, if synergism is widespread in nature, evolution will progress in steps—development will be rapid during the evolution of synergism and then slow down and remain slow until a new crisis occurs, which causes ecosystems to collapse on a wider scale, leading to a new evolutionary step. This agrees with the punctuated equilibria hypothesis, that evolution is concentrated in very rapid events of speciation (Gould and Eldredge, 1977).

Interestingly, the postulated decrease in mutation rates may provide a clue to the mechanism behind mass extinctions. After a long period with stable conditions, organisms in old ecosystems will have low mutation rates. When a crisis occurs, organisms in younger ecosystems will benefit from higher mutation rates and may thereby spread and oust many of the genetically "aged" organisms with lower mutation rates. The observed ecosystem shifts along the Norwegian Skagerrak coast appeared to be related to gradual environmental changes. Consequently, physical, catastrophic events, such as meteorite impacts, may not be a prerequisite for mass extinctions, as ecosystems may collapse as a result of gradually changing environmental conditions, such as increasing or decreasing temperature. If mass extinctions do occur at regular 26-million-year intervals, as suggested by Raup and Sepkoski (1984), this could indicate the length of the genetic "aging process" that renders ecosystems on a wider scale susceptible to collapsing. Alternatively, this genetic aging process could be approximately 62 million years, corresponding to a more recently published cycle in fossil diversity for which no compelling match with known geophysical cycles was evident (Rohde and Muller, 2005).

8.11 THE BALANCE OF NATURE

The balance of nature has been a background assumption since antiquity (Egerton, 1973), and the notion that there is some form of stability is still a fundamentally important aspect of ecology (May, 1981; McCann, 2000). However, Connell and Sausa (1983, p. 808) state that "if a balance of nature exists, it has proved exceedingly difficult to demonstrate." Margalef (1963) and Odum (1971) have both speculated around the process of succession and describe development of an ecosystem as being directed toward some sort of holistic, stable unity, in line with the superorganism concept (see Egerton, 1973). Odum (1971, pp. 252 and 257) states, "The overall strategy [of ecosystem development] is...directed toward achieving as large and diverse an organic structure as is possible within the limits set by the available energy input and the prevailing physical conditions of existence (soil, water, climate and so on)," and continues, asking, "Do mature ecosystems age as organisms do? In other words, after a long period of relative stability or 'adulthood,' do ecosystems again develop unbalanced metabolism and become more vulnerable to diseases and other perturbations?" Odum further states that "the 'strategy' of succession as a short-term process is basically the same as the 'strategy' of long-term evolutionary development of the biosphere." However, as evolution operates on individual organisms or genes (e.g., Krebs and Davies, 1984), statements like this can be nothing but loose speculations if they are not put into an evolutionary context by discussing how such developments are brought about by benefiting individual organisms.

The repeated incidents of bifurcations along the Norwegian Skagerrak coast are strong evidence suggesting that there is indeed a balance of nature. The mechanism underlying this balance in marine pelagic food webs is the evolution of predator-prey synergism, which is an interaction between organisms that enhances the fitness of all involved participants (in other ecosystems there may be different mechanisms that contribute to balance). Hence, in marine ecosystems, evolution appears to be directional toward synergistic systems and dynamically stable states. However, the perception of the "delicate balance of nature" does not seem to apply for marine spring bloom systems. Rather, high resilience in combination with high variability in the plankton community suggest that marine ecosystems are generally robust to environmental and biological perturbations, and the robustness of the system appears unrelated to homeostatic forces that pull the system back into balance after perturbations. The main threat to the integrity of marine spring-blooming systems seems to be environmental changes and human exploitation of marine resources that affect competition in plankton and thus reduce the resilience of the systems.

8.12 CONCLUSION

Based on evaluation of variability enhancing and variability dampening mechanisms, an ecosystem hypothesis for marine spring-blooming systems has been proposed. The essence of this hypothesis is that bottom-up processes in relation to perturbations in physical and chemical variables generate variability in the plankton community. Synergism in plankton acts to dampen variability and thus generates substantial autocorrelation in the plankton community and may give rise to quasi-stable states. Variable energy flow patterns in the plankton community affect the survival of fish and invertebrates that depend on planktonic prey during early life stages and thus give rise to recruitment variability. Survival in organisms that are planktivorous during early life stages is thus density-dependent with adequate planktonic prey as the limiting factor. In general, after the critical recruitment phase, neither food nor space limit survival, except in some benthic communities. At this and higher trophic levels, predation (including parasitism and diseases) is the main regulating mechanism, and because of widespread opportunism in marine predators, predation generally acts to dampen variability in marine ecosystems. Consequently, as general phenomena, physical and chemical bottom-up processes generate variability in marine pelagic food webs, whereas predation, parasitism, and diseases act to dampen variability. Fisheries targeting larger fishes will thus induce variability in marine ecosystems.

The proposed ecosystem hypothesis provides a generalized conceptual framework of interactions between organisms in marine spring-blooming ecosystems. However, biological interactions are complex. Therefore, there will probably be exceptions to the generalized picture, for example, by pathogens causing epidemic mortality at relatively low host concentrations and thereby enhancing variability, or catastrophic events such as toxic algal blooms or extreme temperatures that cause mass mortality in organisms and thus generate patterns in marine spring-blooming ecosystems that are not included in the proposed ecosystem model. There are also reports of extraordinary meager fish, which may indicate that food limitations may occur at higher trophic levels, as in Arcto-Norwegian cod in 1903 (Hjort, 1914). However, exceptions should not prevent us from drawing generalized pictures of interactions between organisms, nor should generalized pictures be rejected by exceptions.

References

Aure, J., Danielssen, D., Svendsen, E., 1998. The origin of Skagerrak coastal water off Arendal in relation to variations in nutrient concentrations. ICES J. Mar. Sci. 55, 610–619.

Beaugrand, G., Brander, K.M., Lindley, J.A., Souissi, S., Reid, P.C., 2003. Plankton effect on cod recruitment in the North Sea. Nature 426, 661–664.

Bergstom, C.T., Lachmann, M., 2003. The Red King effect: when the slowest runner wins the coevolutionary race. Proc. Natl. Acad. Sci. USA 100, 593–598.

Berkeley, S.A., Chapman, C., Sogard, S.M., 2004. Maternal age as a determinant of larval growth and survival in a marine fish, Sebastes melanops. Ecology 85, 1258–1264.

Birkeland, C., Dayton, P.K., 2005. The importance in fishery management of leaving the big ones. Trends Ecol. Evol. 20, 356–358.

Carpenter, S.R., Kitchell, J.F., Hodgson, J.R., 1985. Cascading trophic interactions and lake productivity. BioScience 35, 634–638.

Chambers, R.C., 1997. Environmental influences on egg and propagule sizes in marine fishes. In: Chambers, R.C., Trippel, E.A. (Eds.), Early Life History and Recruitment in Fish Populations. Chapman & Hall, London, pp. 62–102.

Chambers, R.C., Leggett, W.C., 1996. Maternal influences on variation in egg sizes in temperate marine fishes. Amer. Zool. 36, 180–196.

Cloern, J.E., Jassby, A.D., 2009. Patterns and scales of phytoplankton variability in estuarine–coastal ecosystems. Estuar. Coasts. 33, 230–241.

Connell, H., Sausa, W.P., 1983. On the evidence needed to judge eocological stability or persistence. Am. Nat. 121, 789–824.

Cury, P., Shannon, L.J., Shin, Y.-J., 2003. The functioning of marine ecosystems. In: Sinclair, M., Valdimarsson, G. (Eds.), Responsible Fisheries in the Marine Ecosystem. CABI Publishing, Cambridge, pp. 103–123.

Cushing, D.H., 1975. Marine Ecology and Fisheries. Cambridge University Press, Cambridge.

Cushing, D.H., 1989. A difference in structure between ecosystems in strongly stratified waters and in those that are only weakly stratified. J. Plankton Res. 11, 1–13.

Cushing, D.H., 1990. Plankton production and year-class strength in fish populations: an update of the match/mismatch hypothesis. Adv. Mar. Biol. 26, 249–293.

Dahl, E., Johannessen, T., 1998. Temporal and spatial variability of phytoplankton and Chl a—lessons from the south coast of Norway and the Skagerrak. ICES J. Mar. Sci. 55, 680–687.

Dannevig, A., 1949. The variation in growth of young codfishes from the Norwegian Skagerrak coast. Fiskeridir. Skr. Ser. Havunders. 9, 1–12.

Dragesund, O., Johannessen, A., Ulltang, Ø., 1997. Variation in migration and abundance of Norwegian spring spawning herring (*Clupea harengus* L.). Sarsia 82, 97–105.

Ducklow, H.W., Steinberg, D.K., Buesseler, K.O., 2001. Upper ocean carbon export and the biological pump. Oceanography 14, 50–58.

Durant, J.M., Hjermann, D.O., Ottersen, G., Stenseth, N.C., 2007. Climate and the match or mismatch between predator requirements and resource availability. Clim. Res. 33, 271–283.

Egerton, F.N., 1973. Changing concepts of the balance of nature. Quart. Rev. Biol. 48, 322–350.

Fogarty, M., 2001. Recruitment of cod and haddock in the North Atlantic: A comparative analysis. ICES J. Mar. Sci. 58, 952–961.

Frigstad, H., Andersen, T., Hessen, D.O., Jeansson, E., Skogen, M., Naustvoll, L.-J., et al., 2013. Long-term trends in carbon, nutrients and stoichiometry in Norwegian coastal waters: evidence of a regime shift. Prog. Oceanogr. 111, 113–124.

Gould, S.J., Eldredge, N., 1977. Punctuated equilibria: the tempo and mode of evolution reconsidered. Paleobiol. 3, 115–151.

Granéli, E., Carlsson, P., Olsson, P., Sundström, B., Granéli, W., Lindahl, O., 1989. From anoxia to fish poisoning: the last 10 years of phytoplankton blooms in Swedish marine waters. In: Cosper, E.M., Bricelj, V.M., Carpenter, E.J. (Eds.), Novel Phytoplankton Blooms, Causes and Impacts of Recurrent Brown Tides and Other Unusual Blooms. Springer-Verlag, New York, pp. 407–428.

Greatbatch, R.J., 2000. The North Atlantic oscillation. Stoch. Env. Res. Risk A. 14, 213–242.

Green, B.S., 2008. Maternal effects in fish populations. Adv. Mar. Biol. 54, 1–105.

Hairston, N.G., Smith, F.E., Slobodkin, L.B., 1960. Community structure, population control, and competition. Am. Nat. 94, 421–425.

Hanski, I., Hansson, L., Henttonen, H., 1991. Specialist predators, generalist predators, and the microtine rodent cycle. J. Anim. Ecol. 353–367.

Hare, S.R., Mantua, N.J., 2000. Empirical evidence for North Pacific regime shifts in 1977 and 1989. Prog. Oceanogr. 47, 103–145.

Hjort, J., 1914. Fluctuations in the great fisheries of northern Europe viewed in the light of biological research. Rapp. P.-v. Réun. Cons. int. Explor. Mer 20, 1–228.

Holling, C.S., 1973. Resilience and stability of ecological systems. Annu. Rev. Ecol. Syst. 4, 385–398.

ICES, 2012. Report of the ICES Advisory Committee 2012. ICES Advice, 2012. Book 9.

Ianora, A., Poulet, S.A., Miralto, A., 2003. The effects of diatoms on copepod reproduction: a review. Phycologia 42, 351–363.

Irigoien, X., Harris, R.P., Verheye, H.M., Joly, P., Runge, J., Starr, M., et al., 2002. Copepod hatching success in marine ecosystems with high diatom concentrations. Nature 419, 387–389.

Jiang, W., Jørgensen, T., 1996. The diet of haddock (*Melanogrammus aeglefinus* L.) in the Barents Sea during the period 1984–1991. ICES J. Mar. Sci. 53, 11–21.

Johannessen, T., Tveite, S., 1989. Influence of various physical environmental factors on 0-group cod recruitment as modelled by partial least-squares regression. Rapp. P.-v. Réun. Cons. int. Explor. Mer 191, 311–318.

Johannessen, T., Dahl, E., Falkenhaug, T., Naustvoll, L.J., 2012. Concurrent recruitment failure in gadoids and changes in the plankton community along the Norwegian Skagerrak coast after 2002. ICES J. Mar. Sci. 69, 795–801.

Kjesbu, O.S., 1989. The spawning activity of cod, *Gadus morhua* L. J. Fish Biol. 34, 195–206.

Knutsen, H., Olsen, E.M., Jorde, P.E., Espeland, S.H., André, C., Stenseth, N.C., 2010. Are low but statistically significant levels of genetic differentiation in marine fishes "biologically meaningful?" A case study of coastal Atlantic cod. Mol. Ecol. 20, 768–783.

Koski, M., 2007. High reproduction of *Calanus finmarchicus* during a diatom-dominated spring bloom. Mar. Biol. 151, 1785–1798.

Krebs, J.R., Davies, N.N. (Eds.), 1984. Behavioural Ecology: An Evolutionary Approach. Sinauer Associates, Sunderland, Massachusetts.

Lee, R.F., Hagen, W., Kattner, G., 2006. Lipid storage in marine zooplankton. Mar. Ecol. Prog. Ser. 307, 273–306.

Leis, J.M., 2007. Behaviour as input for modelling dispersal of fish larvae: behaviour, biogeography, hydrodynamics, ontogeny, physiology and phylogeny meet hydrography. Mar. Ecol. Prog. Ser. 347, 185–193.

Lindahl, O., Hernroth, L., 1988. Large-scale and long-term variations in the zooplankton community of the Gullmar fjord, Sweden, in relation to advective processes. Mar. Ecol. Prog. Ser. 43, 161–171.

Lindahl, O., Perissinotto, R., 1987. Short-term variations in the zooplankton community related to water exchange processes in the Gullmar fjord, Sweden. J. Plankton Res. 9, 1113–1132.

Longhurst, A., 2002. Murphy's law revisited: Longevity as a factor in recruitment to fish populations. Fish. Res. 56, 125–131.

Mackas, D.L., Thomson, R.E., Galbraith, M., 2001. Changes in the zooplankton community of the British Columbia continental margin, 1985–1999, and their covariation with oceanographic conditions. Can. J. Fish. Aquat. Sci. 58, 685–702.

Margalef, R., 1963. On certain unifying principles in ecology. Am. Nat. 97, 357–374.

May, R.M., 1981. Models for two interacting populations. In: May, R.M. (Ed.), Theoretical Ecology. Blackwell Scientific Publications, Oxford, pp. 78–104.

Maynard Smith, J., 1977. The limitation of evolution theory. In: Duncan, R., Weston-Smith, M. (Eds.), The Encyclopedia of Ignorance: Life Sciences and Earth Sciences. Pergamon Press, Oxford, pp. 235–242.

McCann, K.S., 2000. The diversity-stability debate. Nature 405, 228–233.

McQueen, D.J., Johannes, M.R.S., Post, J.R., Stewart, T.J., Lean, D.R.S., 1989. Bottom-up and top-down impacts on freshwater pelagic community structure. Ecol. Monogr. 59, 289–309.

Metzgar, D., Wills, C., 2000. Evidence of the adaptive evolution of mutation rates. Cell 101, 581–584.

Michalsen, K., Johannesen, E., Bogstad, B., 2008. Feeding of mature cod (*Gadus morhua*) on the spawning grounds in Lofoten. ICES J. Mar. Sci. 65, 571–580.

Myers, R.A., 1998. When do environment–recruitment correlations work? Rev. Fish Biol. Fish. 8, 285–305.

Myers, R.A., Cadigan, N.G., 1993. Density-dependent juvenile mortality in marine demersal fish. Can. J. Fish. Aquat. Sci. 50, 1576–1590.

Odum, E.P., 1971. Fundamentals of Ecology. W. B. Saunders, Philadelphia.

Overland, J., Rodionov, S., Minobe, S., Bond, N., 2008. North Pacific regime shifts: Definitions, issues and recent transitions. Prog. Oceanogr. 77, 92–102.

Pechenik, J.A., 1999. On the advantages and disadvantages of larval stages in benthic marine invertebrate life cycles. Mar. Ecol. Prog. Ser. 177, 269–297.

Pierson, J.J., Halsband-Lenk, C., Leising, A.W., 2005. Reproductive success of *Calanus pacificus* during diatom blooms in Dabob Bay, Washington. Prog. Oceanogr. 67, 314–331.

Pondaven, P., Gallinari, M., Chollet, S., Bucciarelli, E., Sarthou, G., Schultes, S., et al., 2007. Grazing-induced changes in cell wall silicification in a marine diatom. Protist 158, 21–28.

Prokopchuk, I., Sentyabov, E., 2006. Diets of herring, mackerel, and blue whiting in the Norwegian Sea in relation to *Calanus finmarchicus* distribution and temperature conditions. ICES J. Mar. Sci. 63, 117–127.

Ratkova, T.N., Wassmann, P., Verity, P.G., Andreassen, I., 1998. Abundance and biomass of pico-, nano-, and microplankton on a transect across Nordvestbanken, north Norwegian shelf. Sarsia 84, 213–225.

Raup, D.M., Sepkoski, J.J., 1984. Periodicity of extinctions in the geologic past. Proc. Natl. Acad. Sci. USA 81, 801–805.

Reigstad, M., Wassmann, P., Ratkova, T., Arashkevich, E., Pasternak, A., Øygarden, S., 2000. Comparison of the springtime vertical export of biogenic matter in three northern Norwegian fjords. Mar. Ecol. Prog. Ser. 201, 73–89.

Ridley, M., 2003. Evolution. Blackwell Scientific Publications, London.

Robison, B.H., Bailey, T.G., 1981. Sinking rates and dissolution of midwater fish fecal matter. Mar. Biol. 65, 135–142.

Rohde, R.A., Muller, R.A., 2005. Cycles in fossil diversity. Nature 434, 208–210.

Rose, K.A., Cowan, J.H., Winemiller, K.O., Myers, R.A., Hilborn, R., 2001. Compensatory density dependence in fish populations: Importance, controversy, understanding and prognosis. Fish Fish. 2, 293–327.

Rothschild, B.J., 1986. Dynamics of Marine Fish Populations. Harvard University Press, Massachusetts.

Saba, G.K., Steinberg, D.K., 2012. Abundance, composition, and sinking rates of fish fecal pellets in the Santa Barbara channel. Sci. Rep. 2 (716), 1–6.

Sabine, C.L., Feely, R.A., Gruber, N., Key, R.M., Lee, K., Bullister, J.L., et al., 2004. The oceanic sink for anthropogenic CO_2. Science 305, 367–371.

Sargent, J.R., Falk-Petersen, S., 1988. The lipid biochemistry of calanoid copepods. Hydrobiologia 167/168, 101–114.

Sarno, B., Glass, C.W., Smith, G.W., 1994. Differences in diet and behaviour of sympatric saithe and pollack in a Scottish sea loch. J. Fish Biol. 45, 1–11.

Schoener, T.W., 1989. Food webs from the small to the large. Ecology. 70, 1559–1589.

Shepherd, J.G., Cushing, D.H., 1980. A mechanism for density-dependent survival of larval fish as the basis of a stock-recruitment relationship. J. Const. int. Explor. Mer 39, 160–167.

Shepherd, J.G., Pope, J.G., Cousens, R.D., 1984. Variations in fish stocks and hypotheses concerning their links with climate. Rapp. P.-v. Réun. Cons. int. Explor. Mer 185, 255–267.

Smayda, T.J., 1998. Patterns of variability characterizing marine phytoplankton, with examples from Narragansett Bay. ICES J. Mar. Sci. 55, 562–573.

Smetacek, V.S., 1985. Role of sinking in diatom life-history cycles: ecological, evolutionary and geological significance. Mar. Biol. 84, 239–251.

Sogard, S.M., 1997. Size-selective mortality in the juvenile stage of teleost fishes: A review. Bull. Mar. Sci. 60, 1129–1157.

Sommer, U., Adrian, R., Domis, L.D., Elser, J.J., Gaedke, U., Ibelings, B., et al., 2012. Beyond the Plankton Ecology Group (PEG) model: mechanisms driving plankton succession. Annu. Rev. Ecol. Evol. Syst. 43, 429–448.

Staresinic, N., Farrington, J., Gagosian, R.B., Clifford, C.H., Hulburt, E.M., 1983. Downward transport of particulate matter in the Peru coastal upwelling: role of the anchoveta, *Engraulis ringens*. In: Suess, E., Thiede, J. (Eds.), Coastal Upwelling: Its Sediment Record. Part A. Responses of the Sedimentary Regime to Present Coastal Upwelling. Plenum, New York, pp. 225–240.

Sverdrup, H.U., 1953. On conditions for the vernal blooming of phytoplankton. J. Cons. int. Explor. Mer 18, 287–295.

Thorson, G., 1950. Reproductive and larval ecology of marine bottom invertebrates. Biol. Rev. 25, 1–45.

Toresen, R., Østvedt, O.J., 2000. Variation in abundance of Norwegian spring spawning herring (*Clupea harengus*, Clupeidae) throughout the 20th century and the influence of climatic fluctuations. Fish Fish. 1, 231–256.

Townsend, D.W., Cammen, L.M., 1988. Potential importance of the timing of spring plankton blooms to benthic-pelagic coupling and recruitment of juvenile demersal fishes. Biol. Oceanogr. 5, 215–228.

Turner, J., 2002. Zooplankton fecal pellets, marine snow and sinking phytoplankton blooms. Aquat. Microb. Ecol. 27, 57–102.

Tveite, S., 1971. Fluctuations in year-class strength of cod and pollack in southeastern Norwegian coastal waters during 1920–1969 Fiskeridir. Skr. Ser. Havunders. 16, 65–76.

Underdal, B., Skulberg, O., Dahl, E., Aune, T., 1989. Disastrous bloom of *Chrysochromulina polylepis* (Prymnesiophyceae) in Norwegian coastal waters 1988—mortality in marine biota. Ambio 18, 265–270.

Ware, D.M., 1975. Relation between egg size, growth, and natural mortality of larval fish. Can. J. Fish. Aquat. Sci. 32, 2503–2512.

Wassmann, P., 1991. Dynamics of primary productivity and sedimentation in shallow fjords and polls of western Norway. Oceanogr. Mar. Biol. Annu. Rev. 29, 87–154.

Wassmann, P., Andreassen, I., Rey, F., 1998. Seasonal variation of nutrients and suspended biomass on a transect across Nordvestbanken, north Norwegian shelf, in 1994. Sarsia 84, 199–212.

Wichard, T., Poulet, S.A., Halsband-Lenk, C., Albaina, A., Harris, R., Liu, D., et al., 2005. Survey of the chemical defence potential of diatoms: screening of fifty species for α,β,γ,δ-unsaturated aldehydes. J. Chem. Ecol. 31, 949–958.

Yasumoto, T., Underdal, B., Aune, T., Hormazabal, V., Skulberg, O.M., Oshima, Y., 1990. Screening for haemolytic and ichtyotoxic components in *Chrysochromulina polylepis* and *Gyrodinium aureolum* from Norwegian waters. In: Granèli, E., Sundström, B., Edler, L., Anderson, D.M. (Eds.), Toxic Marine Phytoplankton. Elsevier, New York, pp. 436–440.

Zagami, G., Badalamenti, F., Guglielmo, L., Manganaro, A., 1996. Short-term variations of the zooplankton community near the Straits of Messina (North-Eastern Sicily): relationships with the hydrodynamic regime. Estuar. Coast. Shelf Sci. 42, 667–681.

Index

Note: Page numbers followed by "*f*" and "*t*" refers to figures and tables respectively.

J

K

L

M

N

O

P

Printed and bound by CPI Group (UK) Ltd, Croydon, CR0 4YY
03/10/2024
01040420-0003